Beautiful China 美丽中国 Ⅱ

住宅景观 2

Residential Landscape

佳图文化 编

华南理工大学出版社
SOUTH CHINA UNIVERSITY OF TECHNOLOGY PRESS
·广州·

图书在版编目（CIP）数据

住宅景观．2：汉英对照 / 佳图文化编．— 广州 ：华南理工大学出版社，2015.5
（美丽中国；第 2 辑）
ISBN 978-7-5623-4472-8

Ⅰ．①住…　Ⅱ．①佳…　Ⅲ．①住宅 – 景观设计 – 作品集 – 中国 – 现代　Ⅳ．① TU241

中国版本图书馆 CIP 数据核字（2014）第 268360 号

美丽中国Ⅱ：住宅景观 2
佳图文化 编

出 版 人：韩中伟
出版发行：华南理工大学出版社
（广州五山华南理工大学 17 号楼，邮编 510640）
http://www.scutpress.com.cn　E-mail: scutc13@scut.edu.cn
营销部电话：020-87113487　87111048（传真）
策划编辑：赖淑华
责任编辑：周　芹
印 刷 者：广州市中天彩色印刷有限公司
开　　本：889mm×1194mm　1/12　印张 :22.5
成品尺寸：285mm×285mm
版　　次：2015 年 5 月第 1 版　2015 年 5 月第 1 次印刷
定　　价：320.00 元（USD 60.00）

Preface 前言

As the sequel of *Beautiful China I*, *Beautiful China II* selected the representative landscape for housing and public business in China and continued to professionally analyze these "beautiful" cases in detail from multiple perspectives. Content layout starts from the aspects of the keywords, features, and design concept, providing a large number of professional technical drawings, with rich and detailed information. Our effort is to give landscape designers and practitioners new vision and inspiration, create more beautiful landscape, and achieve the Chinese dream of constructing beautiful China and realizing the Chinese nation's sustainable development in landscape design.

《美丽中国Ⅱ》是《美丽中国Ⅰ》的延续之作，精选了国内具有代表性的住宅景观以及公共商业景观，继续从多角度详细且专业地分析了这些“美丽”的景观案例。内容编排上，分别从景观案例的关键点、亮点、设计理念等方面入手，配合大量的专业技术图纸，资料丰富而详实。我们的努力是为了给予设计师及景观从业者新的视觉、新的灵感，以创作出更多更美的景观，在景观设计上共同实现建设美丽中国、实现中华民族永续发展的中国梦。

CONTENTS

EUROPEAN STYLE
欧式风格

SOUTHEAST ASIAN STYLE
东南亚风格

目录

MEDITERRANEAN STYLE
地中海风格

ART DECO STYLE
ART DECO 风格

OTHER STYLES
其他风格

European Style

欧式风格

Lush Vegetation
植被浓密

Magnificence
气势宏大

Exquisite Details
细部精美

KEYWORDS 关键词

Atrium Landscape
中庭景观

European Atmosphere
欧洲气息

Venice Waterscape
威尼斯水景

European Style
欧式风格

Location: Shaoxing, Zhejiang
Landscape Design: L & A Design Group
Landscape Area: 73,000 m^2

项目地点：浙江省绍兴市
景观设计：奥雅设计集团
景观面积：73 000 m^2

LANDSEA Green Residential Area, Nanjing
南京朗诗绿色街区

FEATURES 项目亮点

The design has focused on creating a beautiful skyline. By borrowing landscapes from Shizha Lake and creating the atrium landscapes, it allows all units to have beautiful landscape views.

朗诗绿色街区充分考虑城市天际线美观效果、石闸湖自然景观，对内充分利用中庭景观，尽量做到“户户看景、户户优势”资源最大化。

Overview

Buildings of LANDSEA Green Residential Area are designed in modern simple style, looking elegant and fashionable. The development has first combined sci-tech system with natural eco environment to create a comfortable living environment. Located within close proximity to the lake and park, and supported by the planned 60,000 m^2 commercial facilities, it enjoys complete facilities for recreation, entertainment, shopping, F&B, etc.

项目概况

朗诗绿色街区建筑设计风格为现代简约式，整体简洁明朗，紧跟现代潮流。项目首次融合了科技系统与自然生态环境，创造由内而外的舒适环境。与湖毗邻，与公园为伴，更有6万 m^2 的大型商业配套系统规划，休闲、娱乐、购物、餐饮等生活设施一应俱全。

Design Objective

The design has focused on creating a beautiful skyline. By borrowing landscapes from Shizha Lake and creating the atrium landscapes, it allows all units to have beautiful landscape views.

设计目标

朗诗绿色街区充分考虑城市天际线美观效果、石闸湖自然景观，对内充分利用中庭景观，尽量做到“户户看景、户户优势”资源最大化。

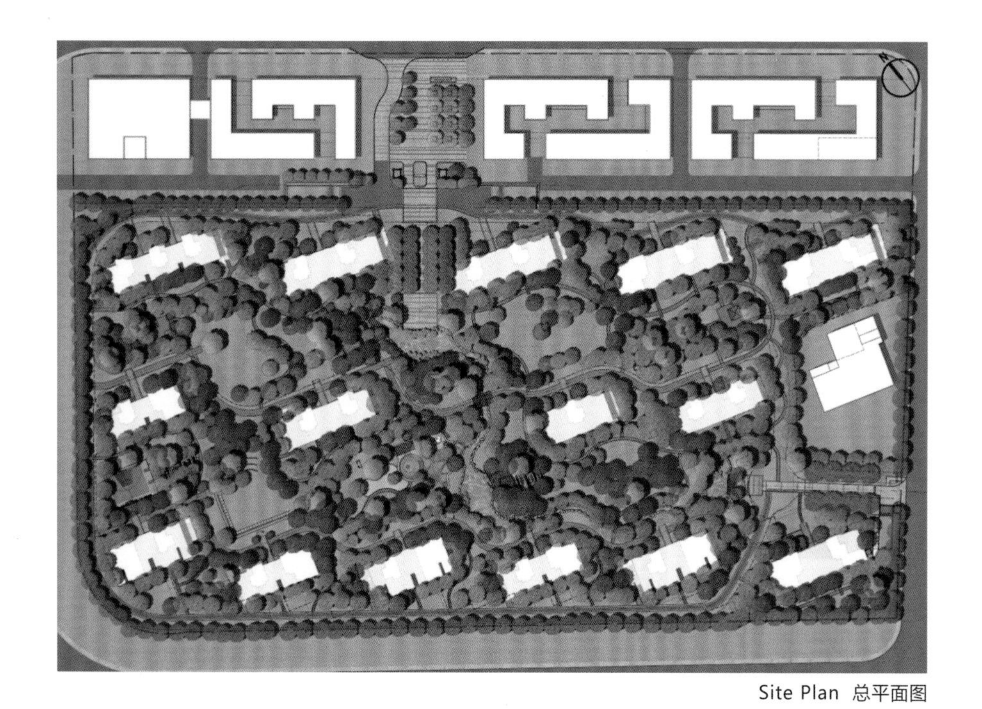

Site Plan 总平面图

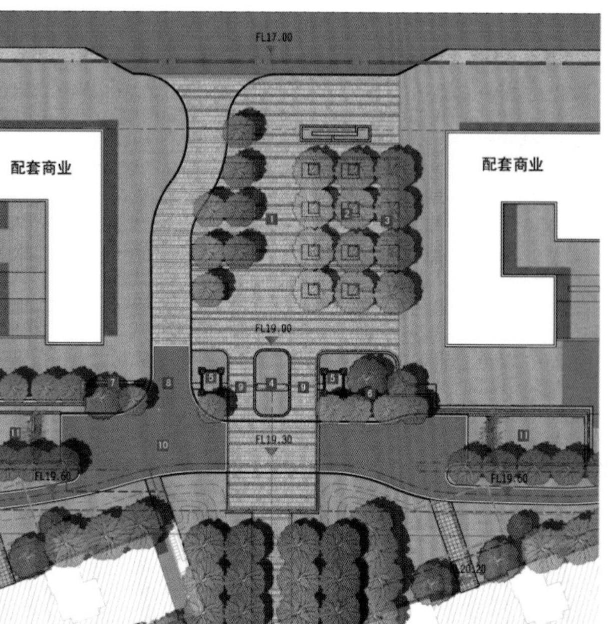

1 商业街铺装广场
2 景观树阵
3 特色景观树
4 LOGO 景石
5 主入口岗亭
6 标准围墙
7 特色景墙
8 车档
9 人行铁艺门
10 景小区车行路
11 地库出入口

Main Entrance Plan 主入口平面图

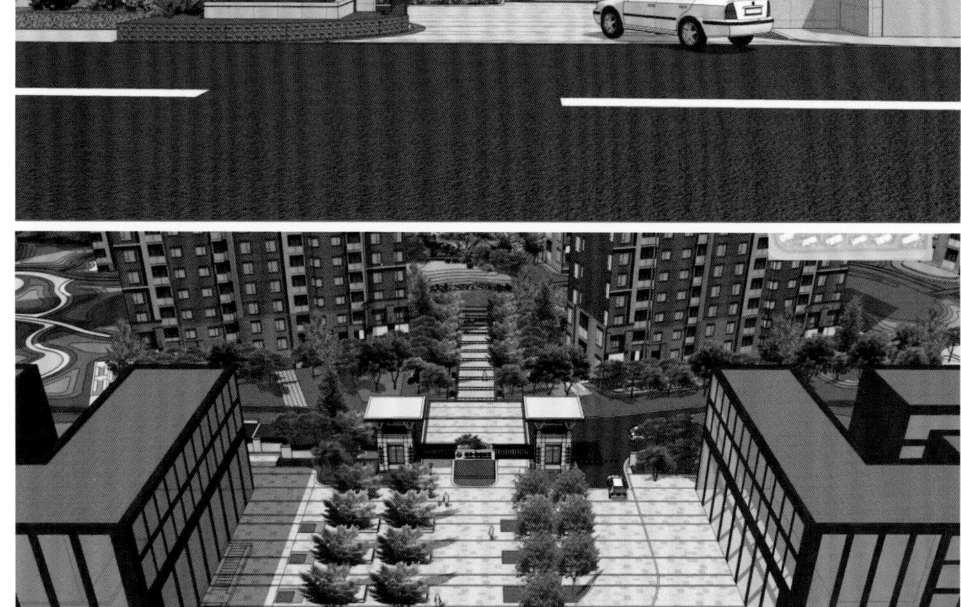

Landscape Planning

1. A 310 m-long central landscape avenue dialogues with the waterscape, reminding people of the Italian Water Towns of the 18th century.

2. A 6,000 m^2 swimming pool with pure blue water in the garden will allow people to enjoy cool summers.

3. Strolling along the 340 m waterfront avenue in the morning or at dusk, one can not only breathe more fresh air but also get closer to water and have good luck.

4. The 11,000 m^2 Victorian central courtyard is unique with classical European-style elegance and harmonious neighborhood atmosphere.

5. There are about 2,000 herbaceous plants of more than 86 species to create a sea of green plants and fragrant flowers.

6. The 5,000 m^2 Orchid Pavilion Garden features zigzag paths and winding streams, providing a best place for people to have a walk.

7. The fantastic 5,500 m^2 Wisteria Garden is ideal for lovers and couples. The drooping wisteria vines with purple blossoms form a magnificent curtain, allowing people to experience the eternal romance when walking by.

景观规划

1. 310 m 的威尼斯中央景观道，水景辉映，使人仿佛置身于 18 世纪的意大利水城。

2. 6 000 m^2 的沁水花园泳池，净蓝池水营造凉爽夏日感。

3. 340 m^2 的滨河景观大道，清晨或黄昏徜徉其中，不仅能呼吸更多清新负离子氧，更能亲水聚财，提升气韵。

4. 11 000 m^2 的维多利亚中心庭院，别具一格，散发着浓浓的古典欧洲尊贵气质与自然和谐的邻里芬芳。

5. 超过 86 种，2 000 株的草本植物，让人进入绿色的海洋，尽情呼吸花草的芬香。

6. 5 000 m^2 的兰亭园是人们休闲散步的好去处，曲径通幽，曲水潺潺，镌绘灵韵，晋风荏苒。

7. 面积 5 500 m^2 的梦幻紫藤苑是甜蜜情侣和温馨夫妇共享二人世界的好去处；长满紫色花瓣的藤蔓一条条垂延而下，交织成一片壮观的紫帘。漫步其下充满恒久浪漫。

KEYWORDS 关键词

Enclosed Structure
围合结构

Seaview Scenery
海景风光

Pleasure and Refreshment
舒适清爽

European Style
欧式风格

Location: Baoting County, Hainan
Landscape Design: SED Landscape Architects Co., Ltd.
Land Area: 36,000 m2

项目地点：海南省保亭县
景观设计：SED 新西林景观国际
占地面积：36 000 m2

Chuangji Nanmei Holiday Phase II, Hainan
海南创基南美假日二期

FEATURES 项目亮点

Blending the concept of "memories" into the project, designers give the development with effect of Spanish legitimate drama, but also attract residents to the carefully designed "journey of Andalusia".

设计方将“回忆”的构思设想融入设计中，不仅赋予了西班牙舞台戏剧的效果，也让居者对设计师极力营造的“安达卢西亚之旅”有种流连不舍的眷恋之情。

Overview

Chuangji Nanmei Holiday Phase II is located in Baoting County, Hainan, which is sitting on the north of Sanya City. It is designed to be enclosed by high-rises that setting along the site, to integrate landscape outside and inside the project together to maximize the landscape resources.

项目概况

海南创基南美假日二期位于海南省保亭县，南邻三亚市。项目整体建筑在总规划中采用周边式布局，以板式高层住宅沿用地围合的结构形式，让优势外部景观资源与小区内部园景相互渗透、融合，使景观资源得到最大化利用。

Design Objective

Blending the concept of "memories" into the project, designers give the development with effect of Spanish legitimate drama, but also attract residents to the carefully designed "journey of Andalusia".

设计目标

设计方将“回忆”的构思设想融入设计中，不仅赋予了西班牙舞台戏剧的效果，也让居者对设计师极力营造的“安达卢西亚之旅”有种流连不舍的眷恋之情。

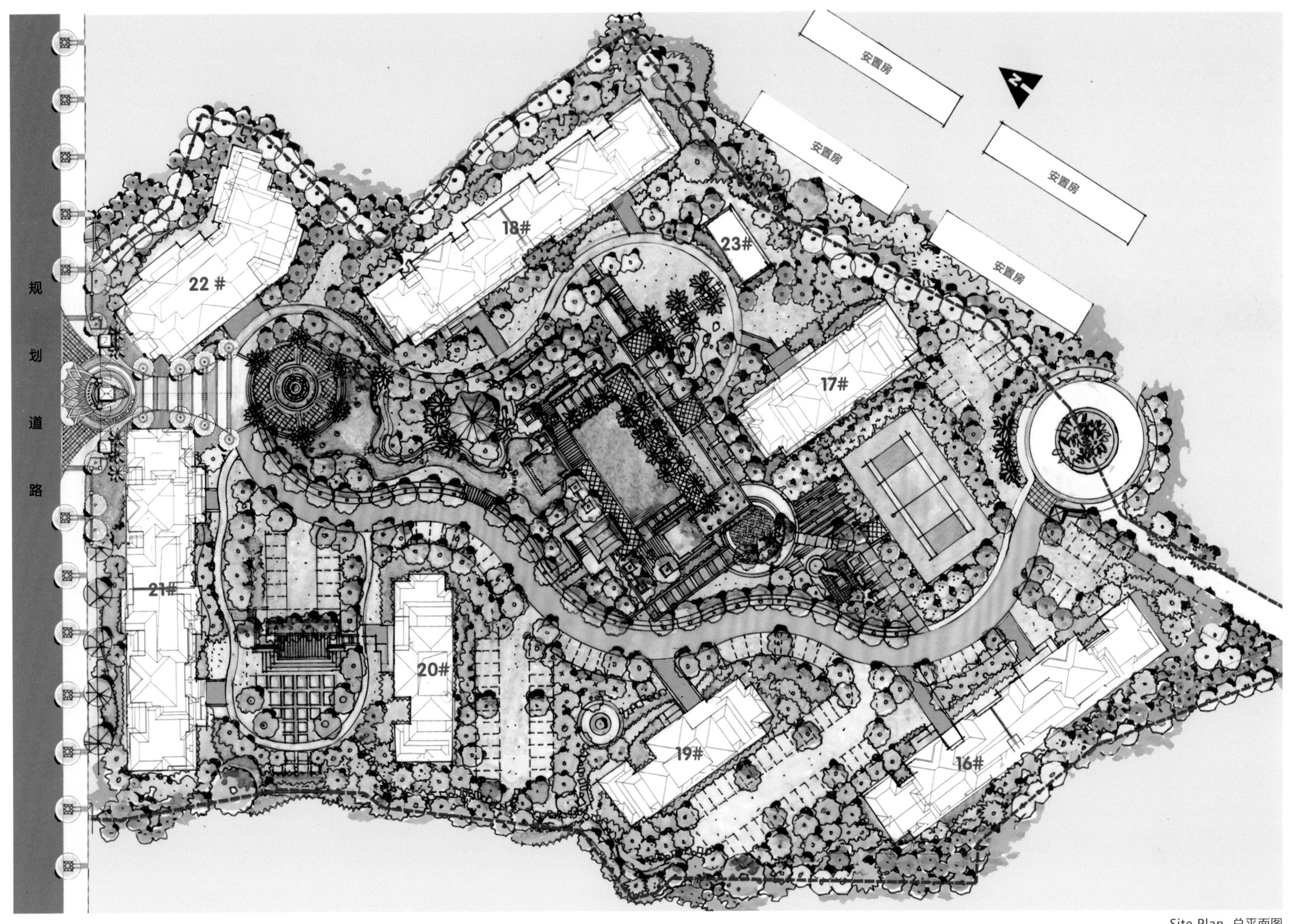

Site Plan 总平面图

Landscape Planning

The landscape design is using Andalusia as the designing source and divided into four groups accordingly based on the composition and functions.

1. Overture — Plaza de Espana of Barcelona

The overture of landscape represents the passionate and enthusiastic Spanish flavor. Warm-toned pavements, featured sculptures and irregular water features of the plaza are working together to create exotic and romantic atmosphere with sunshine and water around. The carefully selected accessories, as well as bright tones with sharp visual impact lead residents into the lively atmosphere of Barcelona to enjoy the artistic flavor and ready for the trip of Andalusia.

2. Main movement — Louisa Park of Costa del Sol, Seville

The main movement of landscape provides large-scale Costa del Sol with silver beach and warm water to offer residents pleasant and refreshment of sunshine coast. Regarded the swimming pool as the core of the landscape group as well as the whole community, landscape designers make full use of terrain height difference to create a cascade to express sense of dynamic and rhyme to the project. Plenty of local plants are adopted, such as palms, and cooperating with grand infinity swimming pool to build "natural town scenery with seaview".

Louisa Park is the largest park of Seville, which is a romantic garden with stream, cascades, potteries, jasmines and Bougainvillea spectabilis Willd in Moorish style. It is designed as an activity zone for adults and children and eager to make it happy with sophisticated supporting facilities to meet the requirements of functions and cooperating tightly with the design theme. Large-scale lawn and plants have ensured high quality environment for recreation and leisure, besides, exquisite feature plants, accessories and pavements are highlighting the landscape environment which is delicate and dynamic.

3. Sub-movement — Ancient city of Cordoba

The sub-movement of landscape is Cordoba, which is an ancient city not far away from Seville and enjoying slow and leisure life.

Divided by the axis road of the project, and compared to other three groups, the landscape group is relatively quiet while designed with leisure spaces of landscape platform and sunken square near the main entrance.

One of the features of the group is the large-sized lawn which has fully represented its elegant planting taste. Sinuous path is running through the whole group so that residents are able to wander in the area to smell the flowers, listen to the birds and enjoy the pleasure. Residents are just like having a holiday in other countries while living in such quiet and pleasant environment.

4. Coda —Palace of the Alhambra

The coda of landscape, Palace of Alhambra is located in a regular town that is Granada but a treasure of Granada as well as Spain. Legendary language is adopted for the group to reach a coda for the residents regarding their psychological feelings. It is connected with phase Ⅰ and making two phases available for both through roundabout. The transition of space is just like travelling across the time and the beginning and end of journey.

景观规划

设计构思以安达卢西亚为设计原点，根据景观空间的组成与功能将其划分为四个组团依次展开。

1. 序曲——巴塞罗那的西班牙广场

景观展开的序曲。热情、奔放的艺术西班牙风情由此呈现。园区休闲广场的暖色调铺装、特色雕塑、不规则水系，营造阳光、海水怀抱的异国浪漫之巅。细致斟酌的小品设计，极具视觉冲击的鲜明色调，尤如置身巴塞罗那的热闹氛围，享受大师艺术气息，使居者宾至如归之余也是体验安达卢西亚之旅的开端。

2. 主乐章——塞维利亚太阳海岸边的路易莎公园

景观展开的主乐章。无垠宽广的“阳光海岸”、银色海滩和暖融融的水域，令游客享受阳光海滨假日的舒适清爽。此组团以泳池区为最大亮点，也是整个小区的核心区域。利用基地高差的特点，设计为大级跌水形式，赋予灵动和韵律感。地域特色种植，如棕榈树等，与大气的无边界泳池打造“自然海景城镇风光”。

塞维利亚最大的花园“路易莎公园”，小溪、瀑布、陶瓷制品以及茉莉花和九重葛处处都带有摩尔式设计风格，是一座充满了浪漫氛围的花园。此区域规划为成人儿童的活动空间，力图营造欢快的氛围。成熟的配套设施丰富了空间的使用功能，动静结合并紧扣设计构思主题。大面积草坪活动区和种植林为居者提供高品质的生活娱乐休闲的环境。特色植物、小品、道路铺装精雕细琢，尤显精巧、活跃的景观情境。

3. 次乐章——科尔多瓦古城

景观展开的间奏。科尔多瓦是一座距离塞维利亚不远的古城。慢节奏的生活气息在此继续延续着。

此组团区域由主要道路将其划分为较独立的空间，与其它三个组团形成动静结合的空间形式。靠近主入口的景观平台与下沉活动广场提供居者赏景、休憩的场所。大面积绿地草坪是此区域的一大特色，优质的种植品位在此体现得淋漓尽致。蜿蜒穿插其中的自然式小径，贯穿整个空间，使居者散步于林间，闻花香、听鸟鸣，识风情。如此闲静的居住环境，仿佛怀着度假散心的闲情雅致身临异国之乡。

4. 尾声——阿兰布拉宫

景观展示的尾声。阿兰布拉宫位于格拉纳达这个平淡无奇的小镇，然而此座宫殿却是当地也是西班牙的瑰宝，将游记印象的传奇语言运用到设计中，所有心理感受在最后的景观组团得到收合的效果。此区域与一期相通，次入口转盘设计将一二期巧妙衔接。空间的过渡转换，恍如时空的跨越、旅程的终始。

KEYWORDS 关键词

British Garden
英式园林

Green Space
绿色空间

Flower Border
花境造景

European Style
欧式风格

Location: Nanhui District, Shanghai
Landscape Design: Palm Landscape Architecture Co., Ltd.
Design Team: Li Huiyun, Chen Fenxi, Yao Dongjie
Landscape Area (Phase Ⅰ): 50,000 m^2

项目地点：上海市南汇区
景观设计：棕榈园林有限公司上海分公司
设计团队：李惠芸、陈奋熙、姚冬杰
景观面积（一期）：5 万 m^2

Tudor Dynasty
长泰东郊御园

FEATURES 项目亮点

Cooperating with British architectural style and positioning of creating natural landscape, at the corner of sinuous road, except the elements for dotted landscape, landscape designers adopt flower border as features.

结合项目的英式建筑风格以及对小区自然式造景的定位，设计师在曲线型道路的转角处，除了一些起点景作用的景观元素，主要以花境造景为特色。

Overview

Project is located in Hangtou Town, Dapudong with Kangqiao Town and Zhoupu Town on its north, setting in the middle area between Yangshan Deepwater Port and lujiazui and Pudong Airport. It is also close to Hangtou Station of Metro Line 11. It has a gross land area approx. 240,000 m^2, gross floor area of over 300,000 m^2 and plot ratio of 1.2, which is a high-end community comprised of townhouses and high-rises.

项目概况

项目地处大浦东航头镇，北邻康桥、周浦，位于陆家嘴、浦东机场与洋山国际深水港的中间区域，地铁 11 号线在航头有停靠站。项目占地约 24 万多平方米，总建筑面积约 30 多万平方米，容积率 1.2，是集联排别墅与高层住宅的高品质社区。

Design Objective

The development is designed with British townhouses while most of them are combined with eight or ten villas forming high density spatial layout and the facades are designed with strong sense of extension, the spaces leaving for landscape are rather hard and tight. Responding to the spatial characteristics of traditional townhouses, landscape designers are trying to win the initiative from green spaces and flower border.

设计目标

项目小区中建筑为英式联排别墅，主要为八联排、十联排式的高密度空间布局，建筑立面延续感较强，而留给景观的空间则相当生硬且局促。针对此类传统联排别墅的空间特质，设计师从绿色空间和花境造景中争取景观的主动权。

Phase 1 Site Plan 一期总平面图

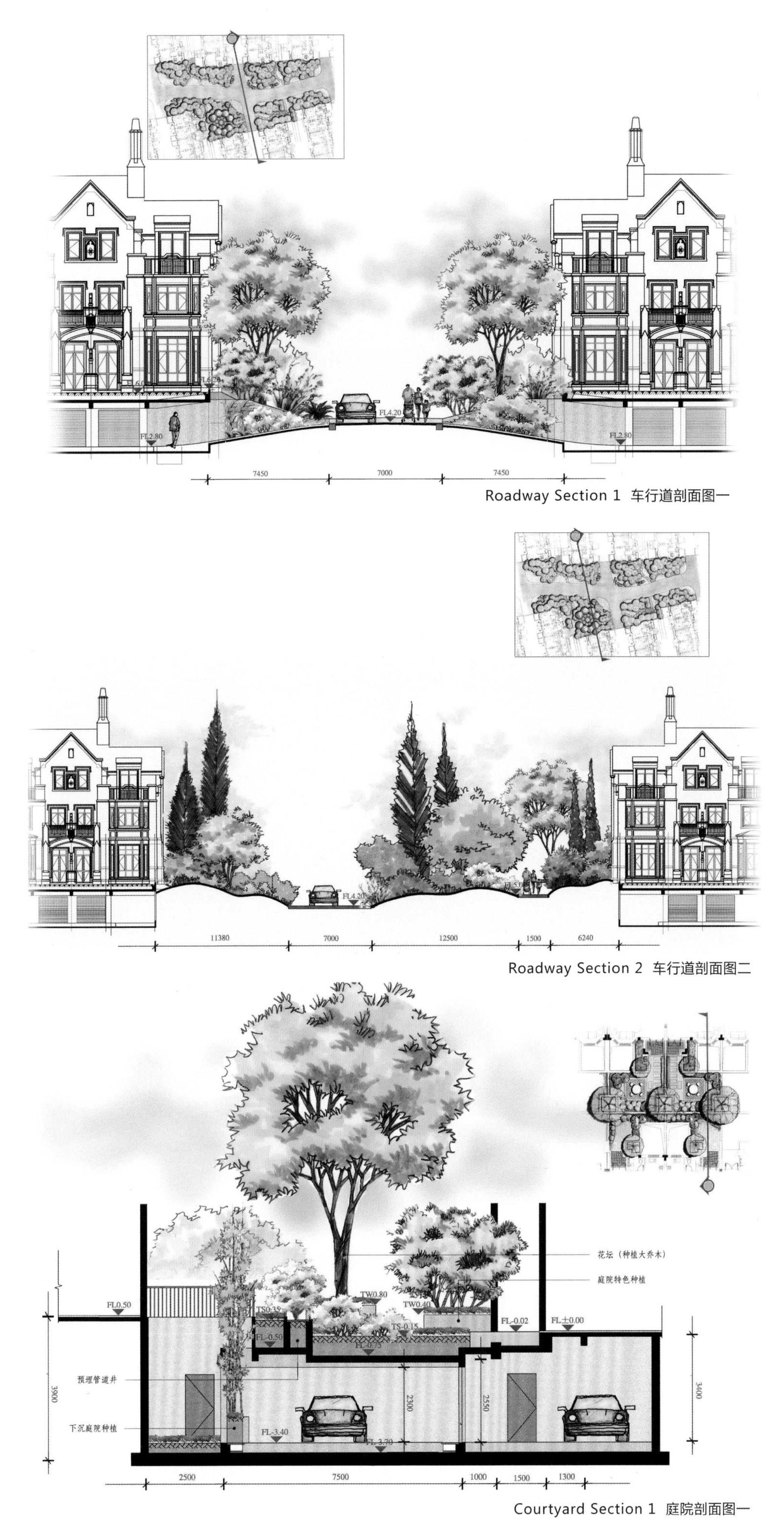

Roadway Section 1 车行道剖面图一

Roadway Section 2 车行道剖面图二

Courtyard Section 1 庭院剖面图一

Landscape Planning

1. Green Space

Despite the spaces that are simple and tight divided by high density townhouses, the rest of these are called green spaces which are paying attention to soften the lines of spaces with the help of plants; to add fun and beauty to spaces by adopting colorful and multi-layered plants; to reduce the exposed massive volume through rich facade of plants. Meanwhile, green spaces are also echoing with the characteristics of nature, closeness and simplicity of British gardens.

For the landscape in this tight space, it should reduce the amount of hard landscaping and overlap of landscape accessories, placing all the elements reasonably from the point of "dotted landscape", especially for the small spaces that people take much more time to view such as paths and gardens leading to the interiors. Plenty of interesting details are added to the project, including the selection of plants, plants with smaller leaves and branches are carefully chosen to contrast against the "large" spaces; besides, colorful plants are used to increase layer to highlight the depth of spaces.

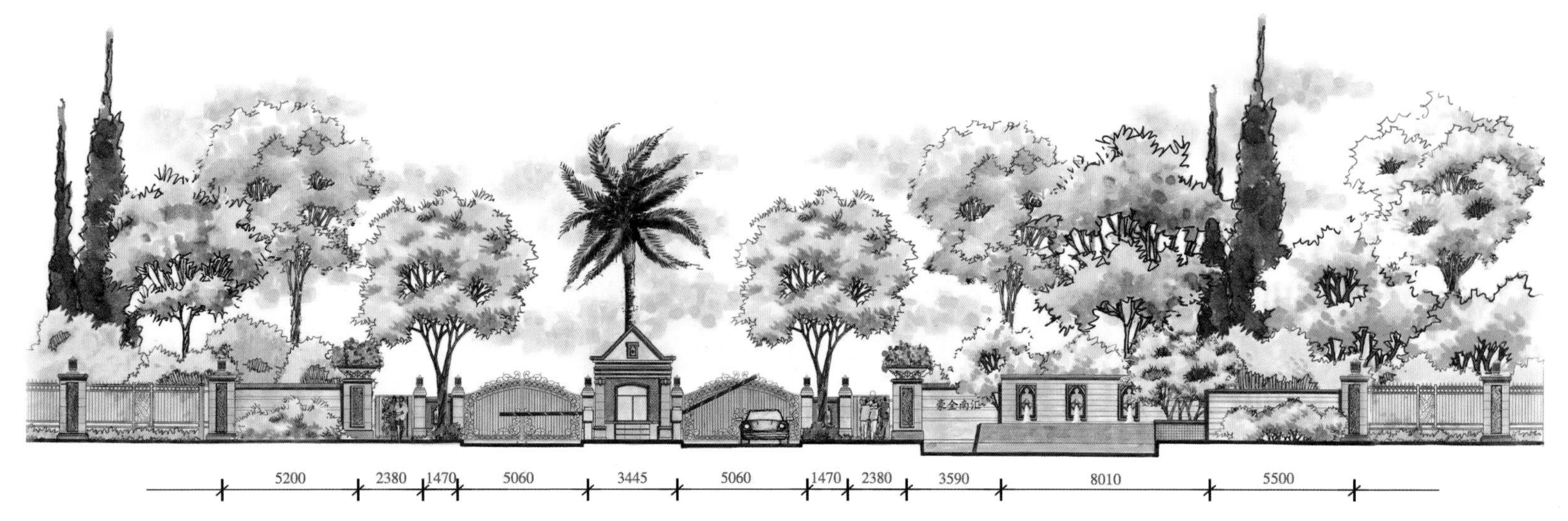

Main Entrance Elevation 主入口立面图

Model Area Plan 样板区平面图

2. Flower Border

Cooperating with British architectural style and positioning of creating natural landscape, at the corner of sinuous road, except the elements for dotted landscape, landscape designers adopt flower border as features. While the word "flower border" exactly comes from the British. Landscape designers learn from the situation of various kinds of wild flowers planting in the borderland of woodland. Studying the plants of flower border regarding layer, height, color, shape, texture and volume, designers place various kinds of perennial flowers in patches mixing planting on the stripped artificial grow beds that in different shapes, to form persistent and multi-layer plant landscape which is with different flowering phases.

景观规划

1. 绿色空间

绿色空间是相对于项目中被高密度的联排式别墅所分割的这种局促且单一、重复的空间，主要强调的是通过植物的围合来柔化空间的线条；通过植物的丰富色彩与层次增加单一重复空间的情趣与美感；通过植物的丰富立面来适当遮掩建筑的体量感。同时，这种绿色空间本身又是与英式园林讲究自然、亲近与纯朴的风格相适应的。

这种相对局促的景观空间里，设计尽量减少硬质园建及小品的堆砌，从"点景物"的角度来考虑各元素的合理搭配。特别是对于入户园路及入户花园此类景观的"末梢"，这是人们使用与观赏时间最多的小空间。在设计中加入了更多耐人寻味的细节，甚至包括在植物的选材上也是有针对性地选择质感较细腻（叶片与株型偏小）的植物品种来反衬空间的"大"；选择色彩较丰富的植物品种营造丰富层次感来突显空间的进深。

2. 花境造景

结合项目的英式建筑风格以及对小区自然式造景的定位，设计师在曲线型道路的转角处，除了一些起点景作用的景观元素，主要以花境造景为特色。而"花境"一词正是源于英国。首先，设计师通过参考自然界中林地边缘地带多种野生花卉交错生长的形态作为本花境设计的原型。在组成花境植物的平面、高度、色彩、形态、质感及体量的统筹设计下，将以多年生花卉为主的不同植物以斑块式，在形状不一的带状人工种植床上进行自然式混交，最终形成层次上高低错落、平面上步移景、花期上次第开放的持久性植物景观。

KEYWORDS 关键词

Dignity and Grand
尊贵大气

Natural and Pleasant
自然舒适

Delicate and Elegant
精致优雅

European Style
欧式风格

Location: Wuzhong Development Zone, Suzhou, Jiangsu
Developer: Ruilong (Suzhou) Real Estate Co., Ltd.
Landscape Design: L&A Design Group
Gross Land Area: 125,050 m^2

项目地点：江苏省苏州市吴中开发区
开发商：苏州瑞龙地产发展有限公司
景观设计：奥雅设计集团
总占地面积：125 050 m^2

Suzhou Wharf Bellagio
苏州九龙仓碧堤半岛

FEATURES 项目亮点

Two different spaces as elegant and grand exterior space and delicate and private interior space are defined by the landscape design to highlight the differences between two spaces and to improve the villa quality.

景观设计通过对“典雅大气的外部空间、精致私密的内部空间”两种空间的不同定义，树立其特质的反差，提升豪宅的气质。

Overview

Located at west of Yinshan Lake of Huanhu Road, Wuzhong Development Zone, Suzhou, Jiangsu, this project occupies advantaged location while at the south of Suzhou ancient town, west of Suzhou Industrial Park, east of Suzhou National High and New Tech Development Zone and Taihu Lake National Tourist Resort and north of Hangzhou City.

项目概况

项目位于江苏省苏州市吴中开发区环湖路尹山湖西侧，地理位置得天独厚，北依苏州古城区，东连苏州工业园区，西接苏州国家高新技术产业开发区和苏州太湖国家旅游度假区，南望杭州。

Demonstration Area Plan 示范区平面图

Design Objective

With the orientation of high-end residence community, the project collocates the architectural types of high-rise residences, stacked townhouses and townhouses in British Georgian Style with the characteristics of elegant symmetry and concise shape. According to the architectural style, the landscape is designed as British natural landscape style. The design objective of the project is to create a dignity, pleasant and elegant British community, to build a warm and romantic but also full of fun and joy British waterfront manor.

设计目标

项目定位为高端住宅项目，建筑类型分为高层住宅、叠拼住宅以及联排住宅。整个社区的建筑风格为英式乔治亚风格，呈现优雅对称、轮廓简明的特征；景观风格根据建筑的特征，定义为英伦自然风景园风格。设计的目标是营造一个尊贵大气、自然舒适、精致优雅的英式人文社区，一个温馨浪漫且富有生活情趣的英式自然滨水庄园。

Landscape Planning

Two different spaces as elegant and grand exterior space and delicate and private interior space are defined by the landscape design to highlight the difference between two spaces and to improve the villa quality. The grand water feature at the entrance or the magnificent clubhouse has shocked the residents to make them feel like staying in front of the garden of British palace. Furthermore, clubhouse garden and waterfront garden are connected with boulevard, natural garden and courtyard garden to form a dynamic integrated space. Delicate landscape nodes and splendid public group spaces can be seen everywhere of the project, such as elegant entrance square, leisure waterfront garden, private courtyard garden, casual backyard garden and spacious green lawn, all of these have made it possible that people may be able to feel the change of quality from grand and elegant to natural and warm.

景观规划

景观设计通过对“典雅大气的外部空间、精致私密的内部空间”两种空间的不同定义，树立其特质的反差，提升豪宅的气质。从彰显大气的入口水景到气势宏大的会所，均给人以强烈震撼，让人仿佛置身于英式宫廷花园面前。此外，设计巧妙地将会所花园、滨水花园与林荫大道、自然花园及宅间花园有机地联系起来，形成有机的空间整体。尊贵典雅的入口广场，自然休闲的滨水花园，优雅私密的宅间花园，对称休闲的后花园，开敞通透的阳光草坪，处处可见精致的景观节点和精彩的公共组团空间，让行走其中的人能感受到从大气典雅到自然温馨的气质变化。

KEYWORDS 关键词

U-shaped Layout
U 形围合

High and Low
高低错落

Natural Curve
自然曲线

European Style
欧式风格

Location: Qingcheng District, Qingyuan, Guangdong
Landscape Design: Guangzhou Taihe Landscape Design Co., Ltd.
Land Area: 163,376 m^2
Floor Area: 119,709 m^2

项目地点：广东省清远市清城区
景观设计：广州市太合景观设计有限公司
占地面积：163 376 m^2
建筑面积：119 709 m^2

Dongcheng King Mountain, Qingyuan
清远东城御峰

FEATURES 项目亮点

Phase I is characterized by its U-shaped layout: there are four building groups of U shape being connected by a main road.

U 形的建筑规划布局是本项目一期用地的显著特征，由此形成一条主干道连通四个 U 形围合的组团空间的结构形式。

Overview

Located on Fengcheng Avenue of Dongcheng District, King Mountain has made reference to the "Royal Garden", and provids the living facilities such as the hotel-style apartments, the commercial street, the kindergarten, the swimming pool, the basketball court and so on. Occupying a total land area of 160,000 m^2, it is developed in three phases. Phase Ⅰ features a land area of 60,000 m^2 and a total floor area of 120,000 m^2 to be a low-density project.

项目概况

东城御峰位于东城区凤城大道中，以清新“御峰花园”为蓝本，配套有酒店公寓、商业风情街、幼儿园等，小区内还有游泳池、篮球场等运动设施。项目总占地超过 16 万 m^2，分三期开发。其中一期占地约 6 万 m^2，建筑面积仅 12 万 m^2，密度相对比较低。

东城
御峰
KING MOUNTAIN
销售中心

Design Objective

The landscape design features economical efficiency, natural environment, suitable size, participating function, art experience, all-well layout, three-dimensional space, and varied styles. It follows the concept of "people orientation, nature first" to highlight the theme of "new style provides unique experience". Making reference to modern simple European style, it has combined the European style with modern techniques perfectly to make the landscape environment keep harmonious with the buildings.

设计目标

景观设计有经济性的原则、自然化的环境、人性化的尺度、参与性的功能、艺术性的感受、均好性的布局、立体化的空间、多样性的风格等原则。整体环境景观设计遵循“以人为本、师法自然”的人性化园林景观设计理念，突出“新格调引领尊贵体验”的主题，以现代简约欧式风格为蓝本，将欧式风格与现代设计手法完美融合，使景观环境与建筑自身特点相呼应、浑然一体。

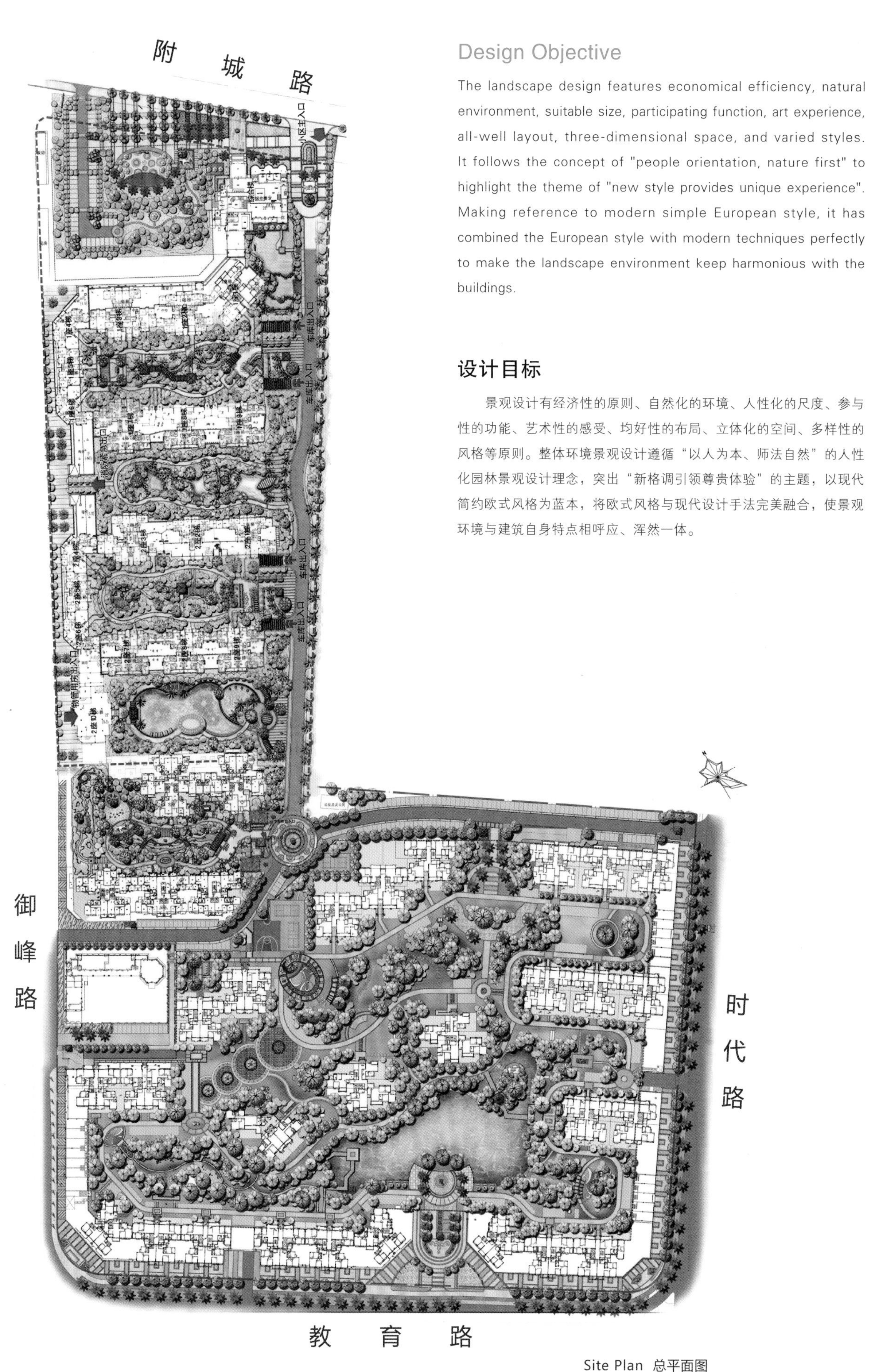

Site Plan 总平面图

御桥
我为园林之雅

Landscape Planning

Landscape Areas: According to the characteristics of the site, it is divided into seven areas: landscape exhibition area, entrance area, leisure area, streams area, children's playground, fitness area, and swimming pool area.

Landscape Elements: It has created multiple visual levels in the limited space. Landscape pavilion, pergola corridor, pavilion and corridor, distinguished pavilion, brooks and cascades, flowing water and waterfalls, fountains and landscape walls, flower pools, tree pools, art sculptures and other landscape items are skillfully designed to create a new-style European garden which looks modern and simple.

Landscape Design of Phase I: In phase I, the buildings are set in four groups which are connected by a main road in U shape. According to the layout of the buildings and the limited space of the underground parking area, the landscape design has created three space levels: private courtyard space, the extension of the private landscape space, and the public landscape space. Thus four building groups are divided into public spaces and half-open spaces. Each group features different landscapes and functions which provide diversified landscape experience.

Composition: The design emphasizes the rhythm of the natural curve to create recreational spaces of different types. Natural curves and soft plant landscapes are combined together. In addition with varied innovative and delicate pavements, corridors, pavilions, waterscapes, vertical flower beds, sculptures, tree pools and sunshine lawn, it has created a brand new living environment in the city.

景观规划

景观分区：根据项目地块特点，主要划分为景观展示区、入口景观区、休闲景观区、溪涧景观区、儿童游乐区、休闲健身区、泳池景观区七大区域。

景观元素：设计在有限的空间里创造出丰富的视觉层次，独具匠心地以观景亭、花架廊、亭廊组合、特色异性亭、溪流跌水、流水瀑布、喷水景墙、跌级花池、特色树池、景观木桥、亲水木平台、休闲园路、汀步、绚丽花溪、阳光草坪、艺术雕塑小品等景观元素来营造具有新格调的现代简约欧式风情园林。

一期景观区设计：U 形的建筑规划布局是本项目一期用地的显著特征，由此形成一条主干道连通四个 U 形围合的组团空间的结构形式，根据建筑规划布局和地下车库顶板的空间限制，在景观规划中，考虑了园区的三个层次空间:住户的私有院落空间；住户的景观空间的延展；住户的公共景园轴空间及其所派生的广场式相适应，并由此将四个组团空间划分为半开放空间和开放空间两种形式。每个组团分别赋予不同的景观特点和功能定位，使之能够提供多样化的景观体验。

设计构图：强调自然曲线的韵律，营造各种高低错落、序列渐进的休闲空间，让自然曲线的构成方式与柔和的生态种植景观和谐统一，配上各种新颖精致的铺地图形，加上各种休闲廊亭、特色水景、立体花池、雕塑小品以及树池、阳光草坪等共同构筑一个全新的都市园林居住环境。

KEYWORDS 关键词

European-style Garden
欧式花园

Landscape Extension
景观延伸

Elegant and Dignified
典雅庄重

European Style
欧式风格

Location: Shaoxing, Zhejiang
Planning and Design: Zhejiang Southeast Architectural Design Co., Ltd.
Landscape Design: Palm Design Co., Ltd.
Land Area: 94,508 m^2
Plot Ratio: 2.0
Green Ratio: 36%

项目地点：浙江省绍兴市
规划设计：浙江东南建筑设计有限公司
景观设计：棕榈设计有限公司
占地面积：94 508 m^2
容积率：2.0
绿地率：36%

The City Star in Shangyu District
上虞城市之星

FEATURES 项目亮点

The community is set close to waters and designers create elevation difference landscape to make liquidity and to show the lively and flexible space of stone, forest, wind, water and light, creating a fresh, sunny city garden.

小区傍水而居，设计师创造高差设计景观，使得景观有流动性，让石、林、风、水、光都展现出生动灵活的意境空间，塑造一个清新、阳光的城市花园。

Overview

The project is located in the north of Shangyu District, its north part is separated with the 3rd Ring Road by a 30 m wide green belt, and three sides of east, west and south are surrounded by planning river, east and south sides are respectively facing Wangshan Road, Wangchong Road across the river, richly endowed by natural landscape. The project covers a total land area of 94,508 m^2 and a total floor area of 189,061 m^2, of which the residential area of 186,886 m^2 and supporting commercial housing area of 2,130 m^2.

Architectural design of the project is based on the neo-classical, style adopting simple and fashionable vertical lines to create light and smooth, simple and elegant visual experience. The landscape cooperating with buildings also adopts abundant classical element composition to form a picture of luxurious and romantic, elegant and dignified European garden.

项目概况

项目坐落于上虞城北，用地北面与城市三环路之间为一道 30 m 宽的城市绿化带隔离，东、西、南三面均被规划河道包围，东、南面分别与望山路、王充路隔河而望，自然景观条件得天独厚。项目用地面积合计 94 508 m^2，地上总建筑面积为 189 061 m^2，其中住宅面积 186 886 m^2，商业与配套用房面积 2 130 m^2。

项目建筑设计在新古典主义的基础上，运用简约朴实而现代时尚的竖向线条营造出轻快流畅及古朴典雅的视觉感受。与建筑相辅相成的景观也大量采用古典元素构图，形成一幅奢华浪漫、典雅庄重的欧式花园的画面。

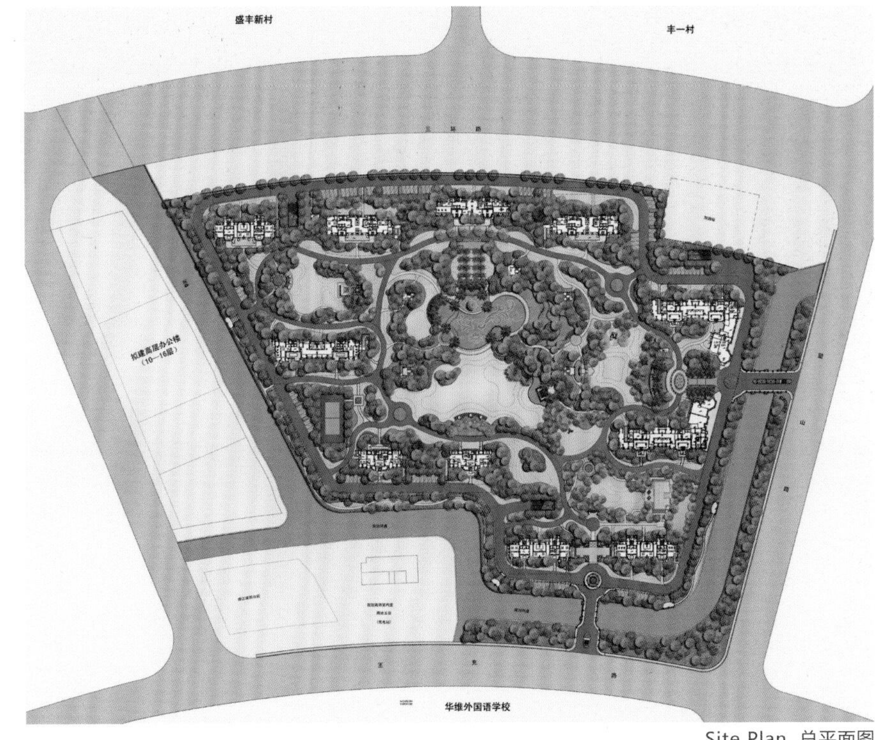

Site Plan 总平面图

Design Objective

Designers want to design a fashionable European city garden where people are harmony with nature, pursuing comfortable and healthy living environment. The whole design is permeated with a kind of healthy concept of "respect for nature, harmonious ecology", creating a fresh, sunshine city garden.

Landscape design of the project focuses on the introduction of ecological and harmonious concept. Designers hope the created landscape combine with buildings according with the surrounding natural environment and avoid excessively emphasizing the landscape itself and breaking the quiet; then enhance the sense of identity of residential tenants to the community through the harmonious ecology; a large amount of green landscape environment space as a place to promote people's emotional exchange, creats the atmosphere to enhance entire community's aesthetic taste.

设计目标

设计师希望设计出一个能够让人与自然和谐共存的时尚欧式城市花园，追求一种舒适与健康的居住环境。整个设计贯穿着一种"尊重自然，和谐生态"的健康理念，塑造一个清新、阳光的城市花园。

项目景观设计的重点是引入一种注重生态和谐的理念。设计方希望营造出来的景观结合建筑融入周围的自然环境，避免突兀地强调景观本身而打破这种宁静；进而通过这种和谐生态，提升住户对小区生活的认同感；设计创建出大量的绿色景观环境空间，作为促进人们情感交流的场所，营造提升整个社区审美情趣的气氛。

Design Concept

The inner of site is level and in irregular shape. Architectural planning's form layout combined "axis and circumference", arranged along lot line from south to north, internal buildings around form an open space.

Central courtyard combined with pool and greenbelt becomes the most important landscape space in axis, refined and elegant. Primary and secondary entrances to the axis adopt specific design techniques to introduce sight line to the central landscape zone. In addition, overhead bottom of residential building forms continuous and transparent landscape gallery to extend landscape inside to the building, which both animate the interior space, and enrich the environment outside the building.

设计构思

场地内部平整，为不规则形状。建筑规划采用"轴线与围合"相结合的布局形式，沿地块边界呈南北向布置，内部建筑围合形成一个开敞空间。

中央庭院结合泳池和绿地，成为小区轴线上最重要的景观空间，精致而高贵。主次入口轴线采用特定的设计手法，将视线引入中央景观区域。除此之外，建筑利用住宅底部架空形成连续通透的景观通道，将景观延伸到建筑内部，既活跃了内部空间，又丰富了外部环境。

Landscape Planning

The main landscape areas within the community include the main entrance area, sub entrance area, pool area, lawn area and riverside landscape belt. The main entrance is with a square to separate the people and vehicles. The annular drop waterscape at the end of the square, collocated with background of plant and foreground of fountain, is clean and momentum, elegant and lively. The secondary entrance is arranged with water wall with strong linear sense and rich layers, and the affiliated carved flowerpot and fountain complement each other. On the axis of the swimming pool is a unique paved square with trees. The tree arrays seem solemn and elegant, highlighting rigorous axis. A regular waterscape comes in the front while going down step by step of the tree arrayed square. The originally quite water surface is embellished with sculptures and European styled lamps. The lawn area is located in the northwest corner of community, surrounded by several groups of graceful buildings. The waterfront landscape uses plant landscaping, different materials and pavement of colors to divide the space, and uses diversified activity places to arrange garden lights and street lights with various shapes.

景观规划

小区内主要景观区域有主入口区、次入口区、泳池区、草坪组团区、沿河景观带。主入口设置了人车分流广场，广场尽头是一座环形跌水水景，配以植物背景和涌泉前景，干净而不失气势，典雅而不失活泼；次入口设置了线条感强烈且层次丰富的喷水景墙，景墙附属的雕塑花钵与涌泉相得益彰；泳池区中轴线上是一个特色铺装林荫广场，成列的树阵庄重优雅，凸显出轴线的严谨，树阵广场拾级而下，迎面是一汪规则式的水景，本来平静的水面上点缀有雕塑和欧式灯具；草坪组团区位于小区西北角，几组姿态优美的建筑将其掬掬围合；滨水景观带利用植物造景和不同材质及颜色的铺装来划分空间，多样化的活动场地中布置着多姿多彩的造型园灯、路灯。

Elevated Floor Design

The bottom floor of residential area is elevated which takes full advantage of design to play a role in traffic, rest and entertainment. There are semi-open children's parks and elderly clubs to provide the residents with direct and convenient resting and communication places. And the designers skillfully use landscaping approaches such as borrowing, framing and hindering to make the interior space extend outside, achieving the effect of increasing space and depth of landscape. Taking the actual situation into account, plants of shade tolerance and strong wind resistance such as Chrysalidocarpus lutescens, Caryota are preferred, and sculptures, landscape walls and other hard landscape are collocated to jointly create a warm, comfortable and funny public activity space.

架空层设计

居住区内楼层底层为架空层。设计充分利用这一特定设计空间，发挥其交通、休息、娱乐的功能，在其中设置了半开放式的儿童兴趣园、老年人俱乐部等，直接为居民提供方便的休息交流场所，并利用特定的场所巧妙运用借景、框景、障景等造园手法，让室内空间向户外延伸，起到了增大空间、加大景深的作用。在植物的选择上也考虑实际情况，多采用耐荫性、抗风性较强的散尾葵、鱼尾葵等，再搭配雕塑、景墙等硬景，共同营造出一个亲切温馨、富有情趣的公共活动空间。

Road System

The clear main road system, in close cooperation with the buildings, smoothly and closely links the major partition. Larger area of spaces, arranged in the main junction of people flow, play a good role in dredging and guiding function. Secondary road system is not stick to one pattern, with diversified form, interesting stools, sculptures and sketchs setted up along the way. All the road systems on both sides reasonably set up unified signs and street lamps, giving full play to their functionality and strengthening the theme for community.

道路系统

主要道路系统与建筑密切配合、明晰了然，将各大分区通达顺畅地紧密联系在一起，在人流主要交汇处均设有较大面积的活动空间，起到良好的疏通和引导功能。次要道路系统不拘一格，形式多样化，并沿途设置别有情趣的坐凳、雕塑和小品。所有道路系统两旁均合理设置风格统一的指示牌和路灯，在充分发挥其功用性的同时亦强化着主题，为小区增色不少。

Plant Arrangement

Plant arrangement conforms to the principle of "planting according to the environment", fully considers the agreement with the architectural style, taking diversity and seasonality into account, matches in multi levels and multi varieties, and combines into the different characteristics of plants community. On the whole there is sparse and dense, high and low, striving to achieve good results in color change and spatial organization.

植物配置

植物配置遵循适地适树的原则，并充分考虑与建筑风格的吻合，兼顾多样性和季节性，进行多层次、多品种搭配，分别组合成特色各异的植物组团群落。整体上有疏有密、有高有低，力求在色彩变化和空间组织上都取得良好的效果。

KEYWORDS 关键词

Center Landscape
中心景观

Model Garden
样板园林

Landscape Node
景观节点

European Style
欧式风格

Location: Shunyi District, Beijing
Owner: China Railway Real Estate Group Co., Ltd.
Landscape Design: United Design Associates, LTD.
Land Area: 125,000 m^2
Landscape Area: 105,000 m^2
Plot Ratio: 1.6

项目地点：北京市顺义区
业　主：中铁置业
景观设计：优地联合（北京）建筑景观设计咨询有限公司
占地面积：12.5 万 m^2
景观面积：10.5 万 m^2
容积率：1.6

Railway Real Estate, Flower County
中铁置业花溪渡售楼处及园区

FEATURES 项目亮点

Utilizing the site's narrow terrain, design team proposed "valley" as the center landscape of the garden through effective communication with the planning and building unit, and connected each building group located in different bottom elevation to form the overall landscape pattern of "enclosure of hillocks and trees, catchment into valley".

设计团队利用场地的狭长地形，经过与规划和建筑单位有效沟通提出园区以“溪谷”为中心景观，串联坐落在不同台底标高上的各个建筑组团，形成“丘林围合、汇水成谷”的整体景观格局。

Overview

The project is positioned as urban high-rise and townhouse community. Different building groups are separated by green space in the community. Landscape design's task is to create community environment of holiday atmosphere in the urban community with relatively higher plot ratio and larger building density, combined with Tuscan intention of building products.

项目概况

本项目定位为城市高层社区和联排别墅。不同建筑组团之间以社区绿地分隔。景观设计任务是在相对容积率较高、建筑密度较大的城市社区里结合建筑产品的意大利托斯卡纳意向营造具有休闲度假气氛的社区环境。

Landscape Analysis Drawing 景观分析图

KEYWORDS 关键词

Magnificent
富丽堂皇

Gorgeous and Elegant
华丽典雅

Eco Greenwood
生态绿林

European Style
欧式风格

Location: Chengdu, Sichuan
Landscape Design: Shenzhen CSC Landscape Engineering Design Co., Ltd.
Land Area: 380,000 m²

项目地点：四川省成都市
景观设计：深圳市赛瑞景观工程设计有限公司
占地面积：38 万 m²

Chengdu Evergrande, Royal Scenic Peninsula
成都恒大御景半岛

FEATURES 项目亮点

Project is positioned as European landscape in the blueprint of France and Italy, perfectly combining the structured atmosphere of French royal gardens and the delicateness and beauty of Italian gardens.

项目定位于以法国和意大利作为蓝本的欧式景观，把法国皇家园林的规整大气和意式园林的精致秀美完美结合。

Overview

Chengdu Evergrande, Royal Scenic Peninsula is the key project delicately built by Evergrande Group, located on the south of Pihe River, Jintang County, Chengdu City, Sichuan Province and the north of Yingbin Avenue, adjacent to the center of Jintang County, with convenient transportation. The plots surrounding natural environment is superior, separated with urban area by a river, with complete living facilities. Project is equipped with business conference center, platinum five-star standard hotel, sports center and large commercial center.

项目概况

成都恒大御景半岛是恒大地产集团精心打造的重点标榜项目，位于四川省成都市金堂县毗河以南，迎宾大道以北，紧邻金堂县中心区，交通便利。地块周边自然环境优越，与城区一河之隔，生活配套设施完备。项目自身配套包括商务会议中心、铂金超五星标准酒店、大型运动中心以及大型商业中心。

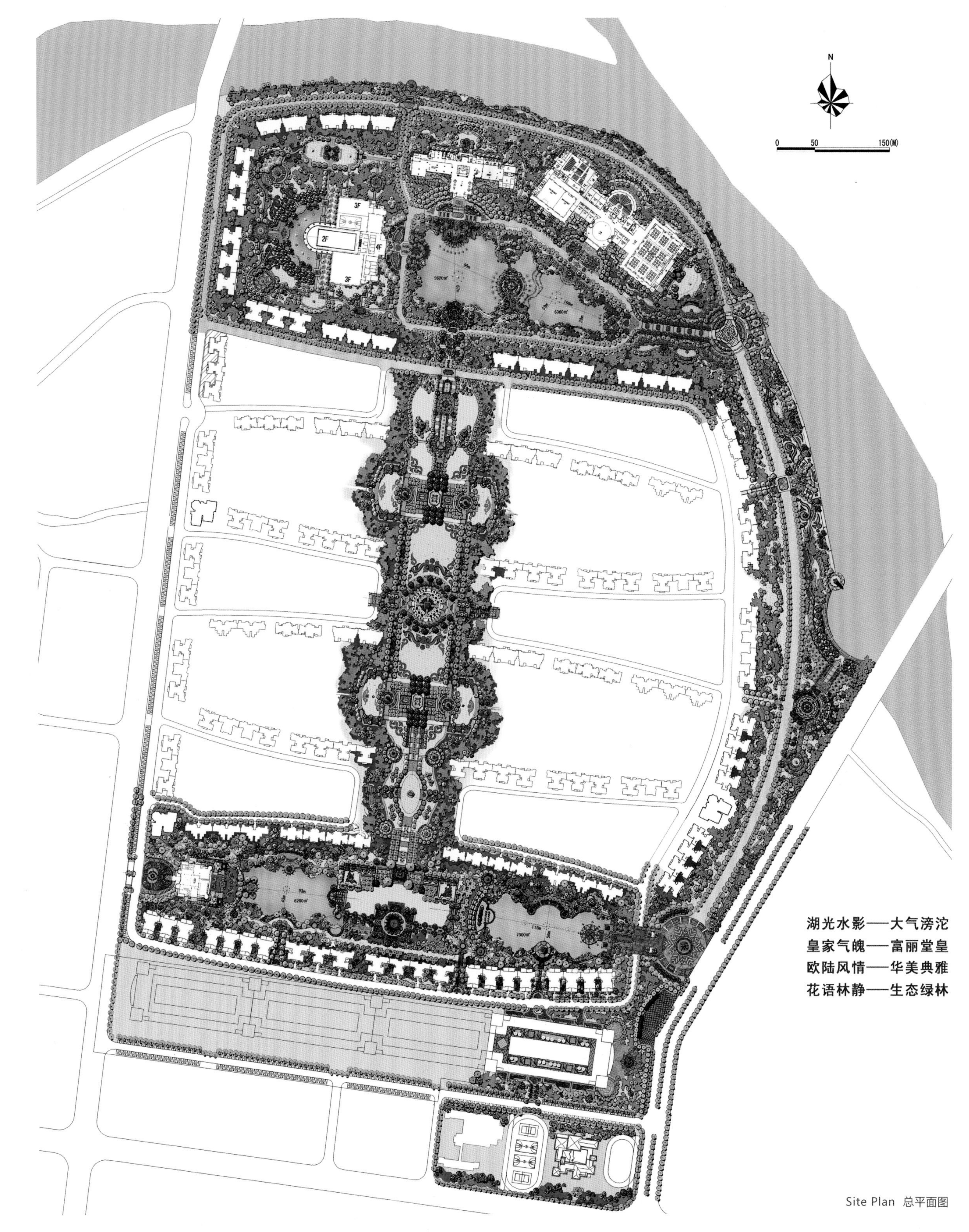

湖光水影——大气滂沱
皇家气魄——富丽堂皇
欧陆风情——华美典雅
花语林静——生态绿林

Site Plan 总平面图

Design Objective

Royal Scenic Peninsula series is one of high-end residential types in Evergrande Real Estate Group and is an European-style residential community of sophisticated planning and low density. The project, located in Jintang County, Chengdu, is positioned as European landscape in the blueprint of France and Italy, and perfectly combines the structured atmosphere of French royal gardens and the delicateness and beauty of Italian gardens, strengthening the region characteristics.

设计目标

御景半岛系列是恒大地产集团的高端住宅类型之一，是高规划起点、低密度的欧陆风情住宅小区。成都御景半岛项目位于金堂县，定位为以法国和意大利为蓝本的欧式景观，把法式皇家园林的规整大气和意大利园林的精致秀美完美结合，并能强化地域特征。

Landscape Planning

With the three landscape lakes and Pihe river as the subject of landscape, multiple landscape nodes combining ornamental value and entertainment are arranged along the water's edge. Project pays attention to create green soft landscape and green fence highlighting the nobleness of Central axes. Eco greenwood in edge cheers the heart and pleases the feeling.

Project is positioned as European landscape in the blueprint of France and Italy, perfectly combines the structured atmosphere of French royal gardens and the delicateness and beauty of Italian gardens, notes the use of detail elements, and strengthens the regional landscape characteristics in the design.

The overall design of plant shows the European style garden. In the plant landscape design, on condition of maintaining consistent core idea of planning and design, the project combines plant with architecture or plant with water landscape, and imitates the nature to take plants as the main view. Considering the influence of Sichuan local characteristics and climate on plant, the project selects the appropriate variety, strengthening axis with regular trees and exquisite parterre and enclosing each space in the natural community way. Through the transformation of different space, the project maximally displays the rich level organically formed by trees, shrubs, and terrain in community, to reach grand, rich and beautiful landscape effect of season changes.

景观规划

以三个景观湖和郫河作为景观主体，沿水边展开设置观赏性和游玩性相结合的多个景观节点。注意绿化软景的营造，造型绿篱衬托中轴的贵气，边缘生态绿林怡情怡性。

项目定位于以法国和意大利为蓝本的欧式景观，把法国皇家园林的规整大气和意式园林的精致秀美完美结合，注意细节元素的运用，在设计中强化地域景观特征。

植物总体设计表现欧陆园林风格。在植物景观设计时，在保持规划与设计的核心理念一致的前提下，或是植物结合建筑、或是植物结合水体造景、或是师法自然以植物为主景，考虑四川当地特色及气候对植物的影响，选择适当品种，以规则式乔木和精致的模纹花坛强化轴线，以自然群落的方式围合各个空间，通过不同空间的转换，最大限度地展示群落中的大树、灌木、地被和地形有机融合形成的丰富层次，达到大气、丰富、四季变化的美丽景观效果。

KEYWORDS 关键词

Five-layer Landscape
五重浓荫

Grapes Valley
葡萄山谷

Venice Rivers
威尼斯水系

European Style
欧式风格

Location: Muping District, Yantai, Shandong
Developer: Longfor (Yantai) Properties Limited
Landscape Design: United Design Associates, LTD.
Land Area: 450,000 m^2

项目地点：山东省烟台市牟平区
开发商：烟台龙湖置业有限公司
景观设计：优地联合（北京）建筑景观设计咨询有限公司
占地面积：45 万 m^2

Yantai Banyan Bay
烟台葡醍海湾

FEATURES 项目亮点

In order to avoid the sense of solid and steep, more than 30,000 grape vines are planted in the valley and decorating with cherry and apple trees, which has been a unique Chinese coastline.

为避免生冷与陡峭的坚硬感，漫山种植三万余株的葡萄树，大量樱桃、苹果树点缀其间，实现中国海岸线的独树一帜。

Overview

Yangma Island and Qianhai are located in the Yellow Sea with location advantage, 9 km far away from north Muping District, Yantai City, and they are also very close to Coast Avenue. In order to make this development integrate with nature, designers make full use of the location advantage and adopt a lot of integration and planning to the project. It has three phases in total: phase Ⅰ is for villas with various architectural shapes such as Seasons Courtyard, Banyan Courtyard and Yihe Villas; phase Ⅱ and Ⅲ are for commercial complex in resort series.

项目概况

养马岛以及前海位于烟台市牟平区城北 9 km 的黄海之中，紧邻滨海大道，区位优势明显。龙湖·葡醍海湾充分运用了地块的天然优势以及后期的整合和规划，力求与天然融为一体。此外，项目共分三期开发：一期以别墅为主，融合了四季小院、葡醍院落、颐和墅等形式；二、三期会突出打造度假产品系中的商业综合体部分。

Design Objective

Combing with advantaged topography, designers employ unique Longfor five-layer landscape series and slope landscape as the core to create Venice rivers and situational gardens which are unusual in north of China.

设计目标

项目结合得天独厚的地理优势，以龙湖独有的五重景观体系和坡地景观为核心，力求打造出北方罕有的威尼斯水系与情境园林。

Site Plan 总平面图

Landscape Planning

Five-layer landscape: when entering into Banyan Bay, residents can only see the plants and flowers while architectures are behind the unique five-layer landscape, and cannot see the end of the roads. The methods, green amount, materials selection and landscape layer are all the lessons to be learnt from.

First layer: megaphanerophyte with height of 7~10 m and diameter over 20 cm;

Second layer: dungarunga and large shrubs with height of 4~5 m;

Third layer: shrubs with height of 2~3 m;

Forth layer: flowers and small shrubs which are the most colorful landscape belt in the garden;

Fifth layer: lawn and ground cover plants.

Venice rivers: Longfor creates the largest artificial nature landscape since last 17 years in this site, to put the project between mountains and villas, also to create the first Venice river system in China. Moreover, a 20,000 m^2 forest park, a 8,000 m^2 borderless swimming pool, a marina, a wetland park and so on are all planned for the development, and project is crossed by the rivers.

Grapes valley: an artificial valley with the height difference of 15 m is created to change the plain topography of Banyan Island. In order to avoid the sense of solid and steep, more than 30,000 grape vines are planted in the valley and decorating with cherry and apple trees, which has been a unique Chinese coastline. Without any doubt, the characteristic of Longfor landscape design is to create landscape by plants and to choose millions of plants to form natural garden community.

Sales Office General Drawing 售楼处总图

景观规划

五重浓荫：走入葡醍海湾，满目密林与花海，建筑掩映在龙湖独有的五重景观中，没有一条道路能够望到尽头。其园林处理手法、绿量、选材与景观层次，都有很强的借鉴意义。

第一重：高 7 ~ 10 m，胸径 20 cm 以上的大乔木；

第二重：小乔木、大灌木（4 ~ 5 m）；

第三重：2 ~ 3 m 高的灌木；

第四重：花卉与小灌木组成园区中色彩最为丰富的景观带；

第五重：草坪、地被。

威尼斯水系：龙湖在葡醍海湾上，建造 17 年来最大的人工自然景观，将中国别墅大成之作，静置山谷林间，并首度呈现威尼斯式运河水系。此外，龙湖充分利用自身天然优势，整体规划 2 万 m^2 海上森林公园、8 000 m^2 无边界泳池及游艇码头、湿地公园等，整个园区内更是水系贯通。

葡萄山谷：葡醍海湾塑造出高差达 15 m 的人工山谷，改变整条海岸线的平缓地貌。为避免生冷与陡峭的坚硬感，漫山种植三万余株的葡萄树，大量樱桃、苹果树点缀其间，实现中国海岸线的独树一帜。植物造景无疑是龙湖产品景观设计的最大特点，甄选成千上万的植物资源，并形成自成体系的园林社区，龙湖堪称一绝。

KEYWORDS 关键词

Varied Layers
层次多变

Classical Charm
古典意趣

Cascades
跌落水景

European Style
欧式风格

Location: Qingdao, Shandong
Developer: Qingdao Hisense Real Estate Co., Ltd.
Landscape Design: Botao Landscape (Australia)
Land Area: 180,000 m^2

项目地点：山东省青岛市
开发商：青岛海信房地产股份有限公司
景观设计：澳大利亚・柏涛景观
占地面积：18 万 m^2

Hisense・Hotspring Dynasty
海信・温泉王朝

FEATURES 项目亮点

The landscape nodes are well arranged. Trimmed arbors, green hedges, axial footpaths and natural stone walls are used to define the patterns of the spaces and allow the buildings to integrate into the surroundings.

景观节点精心布置，利用一些修剪乔木、绿篱、轴线性分布的步道及天然石墙来构筑空间形态，将环境与建筑融为一体。

Overview

Qingdao, situated in the east of Shandong Peninsula, is a major port and economic center in China's eastern coastal region. Wenquan Town of Jimo City is known far and wide for its featured hotspring resources. Hotspring Dynasty is located in the west of Wenquan Town, Jimo City, with Datian Road on the south, the planned Road No. 8 and Road No. 2 on the east and west respectively, and the planned Road No. 1 on the north. It is strategically located at the west portal of Wenquan Town.

项目概况

青岛位于山东半岛东部，是我国东部沿海的主要港口和经济中心。温泉镇作为即墨市下属市镇，其温泉资源以颇具特色而远近闻名。温泉王朝位于即墨市温泉西部，南邻大田路，东西紧邻规划 8 号路和 2 号路，北侧为规划 1 号路，从远景上看，地块位置为温泉镇西部门户。

Site Plan 总平面图

Design Objective

The landscape design for Jimo Hotspring Dynasty, Qingdao has focused on the European landscape style, and employed the modern landscape skills. It is suggested that the key landscape area of the project should use the European landscape design concept and means to make the landscape diversified, interesting and amiable.

设计目标

青岛即墨温泉王朝度假酒店景观设计，以欧洲景观风格为主，同时融入现代景观处理手法。在设计中建议此项目的重点景观区，采用欧洲景观的设计理念和手法，使该项目景观更具多元化、趣味性及亲切感。

Landscape Planning

Main entrance and small square:

The design for this area emphasizes geometric patterns and sequential flowerpots and tree arrays to give the sense of belongings and guidance. The central cascade waterscape has fully considered the visual effect in winter of northern China. The vegetation in the small square is designed in medium height to avoid blocking line of sight, and keep the integrity of the building facade. Moreover, the decorative items and constructions are exquisite and worth carefully appreciating.

Inner courtyard of the business hotel:

The design principle for this part is to create a complete leisure environment for lingering with varied layers and classical charm. The walking routes are smooth and clear, and the landscape nodes are well arranged. Trimmed arbors, green hedges, axial footpaths and natural stone walls are used to define the patterns of the spaces, and allow the buildings to integrate into the surroundings. Since the hotel is open all year round, the environment should be full of vigor in four seasons. So, in the center there is a fountain of classical style to dialogue with a tall pavilion. The arbors are planted on two sides to echo the reflection pool. The background of the pavilion with twining Chinese wistaria has enhanced the romantic atmosphere.

Inner courtyard of the dinning area:

The landscape corresponds to the exterior windows of the buildings and is divided into several parts. The setting of different decorations inside enables visitors to enjoy the outside landscape and experience the extension of the space. They also form the ecological background of the interior spaces.

Townhouse area:

The landscape here seems to continue the landscape of the inner courtyard in the south of the hotel. Thus when overlooking the landscape space, it will present a complete image. The design of the townhouse pays attention to the details and the combination of functions between different residential areas. In the place opposite to the villas, the designers have arranged structures and plants to ensure the privacy of the ground floor without blocking the sunlight. Delicate building groups are both independent and interconnected with small water features to activate the spaces. Small lawns, benches, small sculptures and vines emphasize the romance of the classical countryside style. The treatment of the micro-topography not only softens the boarders of the site but also forms natural landscape belts. The classical-style wall combines the flower pool to form the landscape node which dialogues with the tree arrays to create a friendly, comfortable and beautiful environment.

景观规划

主入口及小广场：

该地段设计中强调几何锦绣纹样与强化序列感的花钵、树阵，旨在强调引导人们的归属感与引导性。中心的跌落水景，充分考虑到北方冬季水景观不能正常运作时的正常效果。小广场的植被不宜过高，避免遮挡视线，保持建筑立面的效果完整，小品与构筑物精细耐看。

商务酒店内庭：

设计师在此处的设计原则是创造一个完整可流连的休闲环境，层次多变，富于古典意趣，人行路线流畅、清晰。景观节点精心布置，利用一些修剪乔木、绿篱、轴线性分布的步道及天然石墙来构筑空间形态，将环境与建筑融为一体。由于酒店作为一个全年对外开放的空间，其环境布置要求四季充满活力。在中心地带设置古典式喷泉，对景布置高凉亭，两侧列植乔木，结合倒影水池呼应成趣。凉亭背景以紫藤缠绕来强调浪漫的景观氛围。

餐饮区内庭：

景观与建筑外窗相对应，分成几个单元，内部不同的小品设置使观者能透过窗外的景观，感受到空间延续，同时又是室内空间的生态背景。

联排别墅：

此处景观与南面酒店的内庭的景观有整体上的连续感，俯瞰整个景观空间时没有断裂感，联排别墅区的设计强调小区间的细节与功能性结合，与别墅对景的位置，设计师均安排了构筑物和植栽，这样充分保护了底层业主的隐私与私密，但又不会对阳光有太多的遮挡。精巧细致的小区之间看似独立又内有联系，小水景的设置可以活泼空间气氛，增加灵动气息。小草坪、长椅、小雕像以及藤蔓植物强调了浪漫的古典田园风貌。微地形的处理既可以软化地块的边缘线，也形成了自然边际景观带。节点的设计是通过古典风格的景墙与花池构成视觉焦点，以树阵为背景，相互依存，共同营造了一个亲切宜人而又美观精致的景观环境。

KEYWORDS 关键词

Elegant and Comfortable
精致舒适

Landscape Garden
景观庭院

Italian Flavor
意大利风情

European Style
欧式风格

Location: Chengdu, Sichuan
Developer: Chengdu Huayi Property Co., Ltd.
Landscape Design: Metro Studio (Italy)
Landscape Area: 75,000 m²

项目地点：四川省成都市
开发商：成都华邑房地产开发有限公司
景观设计：意大利迈丘设计事务所
景观面积：7.5 万 m²

Chengdu Huayi, Live in Sunshine
成都华邑·阳光里

FEATURES 项目亮点

Clean and neat small pieces of courtyard build modern, natural, low-key and simple living space. Italian contemporary art landscape axis running among the space, intends to infuse the wisdom accumulated by art into life.

干净整洁的小庭院体块，构建出一个现代、自然、低调、简约的居住空间。贯穿其间的意大利当代艺术景观轴，意在生活中融入与积累艺术的智慧。

Overview

The project is located in the Pixian County, northwestern suburb of Chengdu City, between the Ring Expressway and the 3rd Ring Road. The plot is flat, overlooking the Xihu Lake Park in the east, in the west there being city green space in the planning, northwest to the famous Southwest Jiaotong University, with the total landscape area of 75,000 m² and the landscape area of 10,756 m² in display area. The project includes display area, phase Ⅰ and phase Ⅱ, planning building layout of "one center, two axes and four groups". The north is high-rise area and south is Western-style residential area. The whole building overhead about 5 m, is equipped with two-level underground garage.

项目概况

项目位于成都市西北郊郫县境内，绕城高速与三环路之间；地块方正平坦，东侧可远眺犀湖公园，西侧有规划中的城市绿地，西北与著名学府西南交大相望，总景观面积 7.5 万 m²，展示区景观面积 10 756 m²。项目包括展示区与一、二期，规划为“一心两轴四组团”的建筑布局，北面为高层区，南面为洋房区。建筑整体架空约 5 m，设置有两层地下车库。

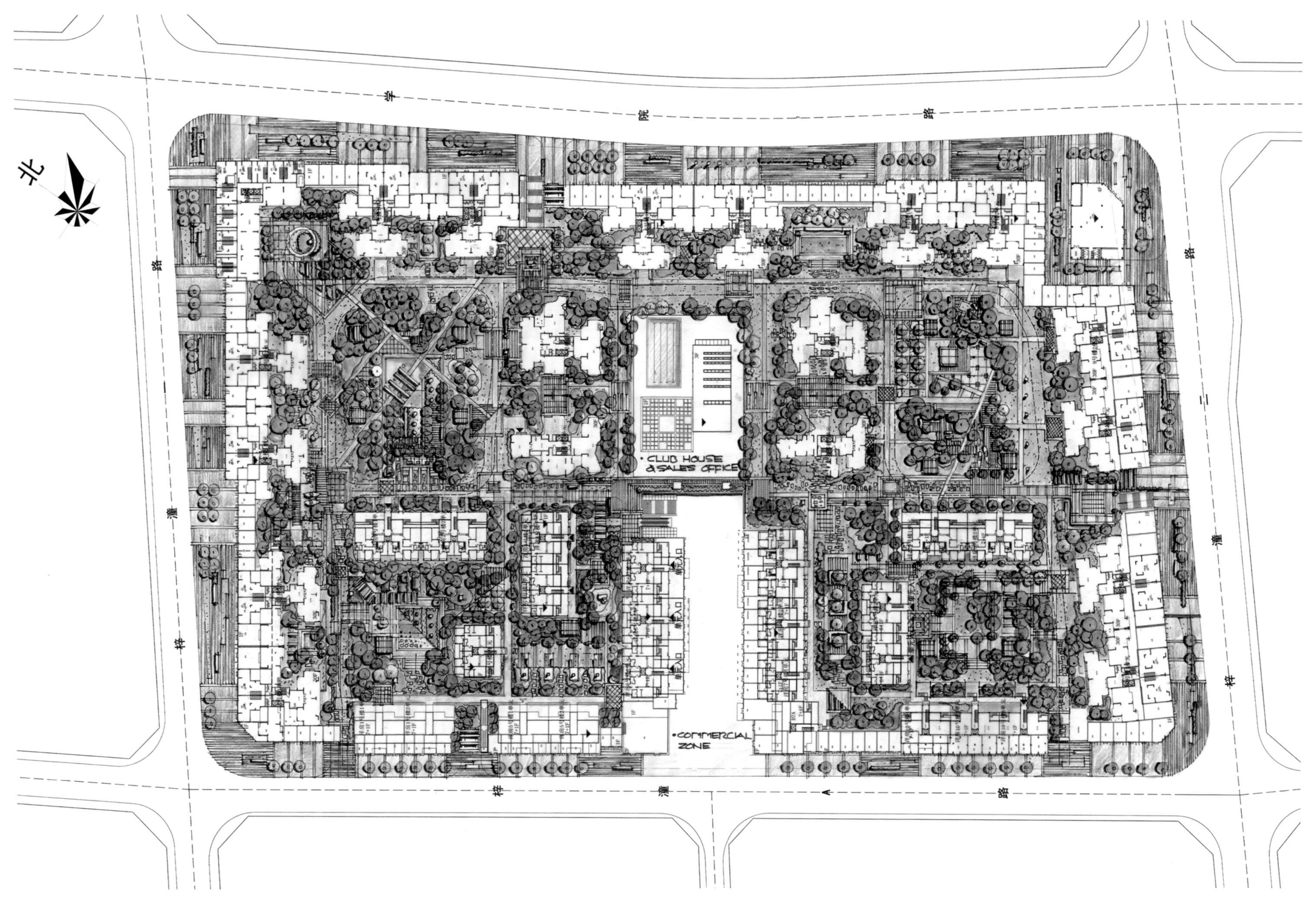

Site Plan 总平面图

Design Objective

In order to highlight the "sunshine" residential theme, the project specially selected Italian town as design template, integrated the theme of "health, art, communication" into the landscape to echo with the building facade in bright colors, and narrated the minimal art and romance of Italy with concise and modern geometric design technique. Making full use of the space, color, lighting changes and other landscape elements, together with enclosed space of hard and soft elements, created different landscape experience. In the environmental design details, conciseness and vividness embody the distinct features of the modern life, at the same time collocated with the sunken style business district, clearing territory feeling and place feeling, to express individuality with the clear lines and clear composition.

设计目标

为了凸显"阳光里"住宅主题，特选取意大利风情小镇作为设计模板，将"健康、艺术、互动"的主题融入到景观中，呼应色彩明快的建筑立面，以简洁现代的几何设计手法与元素，叙述意大利的质朴艺术与浪漫风情。充分利用空间、颜色、光线的变化等景观要素，及硬质和软质元素围合空间，创造出不同的景观体验。在环境细节设计上，简洁明快体现鲜明的现代生活特征，同时配合下沉式商业区，明确领域感与场所感，以清晰的线条与简练的构图来彰显个性。

Landscape Planning

Display area concentrated whole landscape concept of future residential community to show well-built landscape avenue, art gallery, water garden, lawn and other comprehensive landscape effect. By creating landscape environment, comfortable environment, various color, beautiful plants, delicate creations and artistic atmosphere bring deep experience to customers. After difference consideration of permanent landscape avenue and temporary landscape in sample houses area, the effect of landscape are satisfied on condition of controlling the construction cost and difficulty of temporary landscape. The landscape avenue in main entrance highlights logo and guiding function, refining landscape design elements from the building facade and signifying the exclusive symbol of plots used in waterscape pattern and creations; combining with planting, landscape is endowed by sense of rhythm and movement. Landscape garden space is displayed in blending group style, looking good in dynamic or static state; art gallery area emphasizes the interactive and artistic atmosphere.

Landscape design project of the phase Ⅰ and phase Ⅱ extracts comfortable and ecological language from the nature, adds curve lines into linear composition, and the residential area is named after the Italian towns. Clean and neat small pieces of courtyard build a modern, natural, low-key and simple living space. Contemporary Italian artistic landscape axis running among the space, inspired by Italian Renaissance art achievements and contributions, intends to infuse the wisdom accumulated by art into life. As a crowded landscape platform, it uses the creating techniques of space changes, art insert and decorative artwork to extend art axis to east and west sides from the sales center, highlighting the high-quality life realm. Among them, the scenic wall, as the center of art axis, is rich in Italian literature flavor, with the function of pointing out the scenery and appreciating. Slanting road design breaks the routine, also reflects the impact force and aesthetic feeling of lines and plane intersecting in modern style, in order to facilitate the households freely passes.

景观规划

展示区浓缩未来小区整体景观概念，呈现出精心打造后的景观大道、艺术长廊、水苑和坪原等综合景观效果。通过景观环境的营造，以舒适的环境、多样的色彩、绚丽的植物、精致的小品、艺术的氛围，给客户带来深刻体验。区别考虑永久保留的景观大道和临时性的样板房区景观，在满足景观设计效果的同时控制临时性景观的造价、施工困难度。主入口景观大道突出标识与引导功能，从建筑立面提炼景观设计的元素，并将地块的专属标识符号化，运用于水景格局与小品的营造；配合绿化种植，赋予节奏感和律动感。展示景观庭院空间，组团式庭院相互交融，宜动宜静；艺术长廊区域强调其互动性与艺术氛围。

项目一、二期的景观设计从自然提取舒适生态的语言，在直线构图里加入曲线的形态，并借以意大利风情小镇命名住宅区域。干净整洁的小庭院体块，构建出一个现代、自然、低调、简约的居住空间。贯穿其间的意大利当代艺术景观轴，灵感来自于意大利文艺复兴的艺术成果与贡献，意在生活中融入与积累艺术的智慧。作为一个聚集人群的景观平台，在营造手法上利用空间的变化、艺术的穿插、艺术品的点缀，使艺术轴由售楼中心延伸至东西两边，突显高品质的生活境界。其中，景墙作为艺术轴的中心，富有浓厚的意大利文艺气息，具有点景与观赏功能。打破常规的斜线道路设计，为了方便住户间人流通行，也体现了现代风格的线面相错的冲击力与美感。

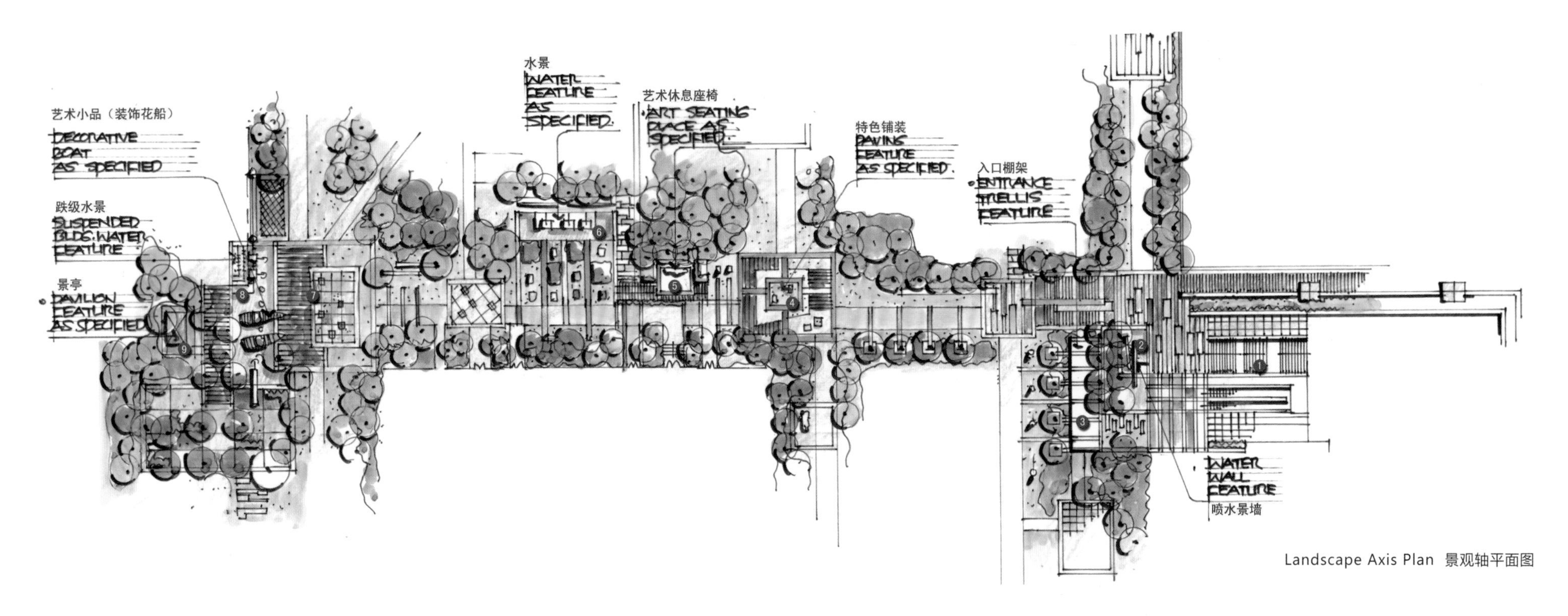

Landscape Axis Plan 景观轴平面图

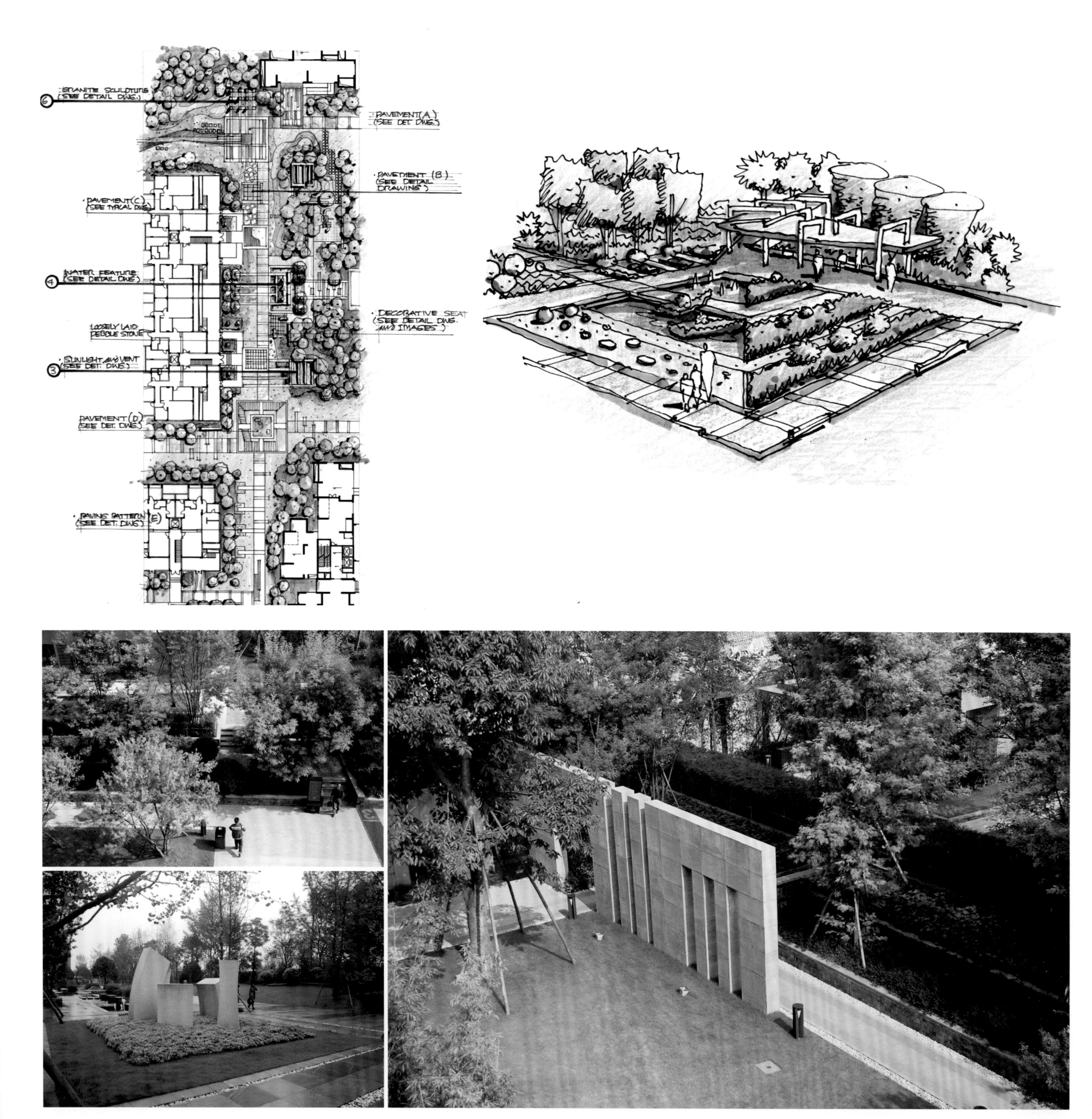
GRANITE SCULPTURE (SEE DETAIL DWG.)
6
PAVEMENT(A) (SEE DET. DWG.)
PAVEMENT (B) (SEE DETAIL DRAWING)
PAVEMENT(C) (SEE TYPICAL DWG.)
WATER FEATURE (SEE DETAIL DWG.)
4
DECORATIVE SEAT (SEE DETAIL DWG. AND IMAGES)
LOOSELY LAID PEBBLE STONE
SUNLIGHT AND VENT (SEE DET. DWG.)
3
PAVEMENT(D) (SEE DET. DWG.)
PAVING PATTERN (E) (SEE DET. DWG.)

KEYWORDS 关键词

Small Swiss Town
瑞士小镇

Place of Peace
世外城邦

Stacked Rocks
山石叠砌

European Style
欧式风格

Location: Bao'an District, Shenzhen, Guangdong
Developer: Shenzhen Hwaloilee Group
Landscape Design: Botao Landscape (Australia)
Landscape Area: 60,000 m^2

项目地点：广东省深圳市宝安区
开发商：深圳市华来利实业公司
景观设计：澳大利亚·柏涛景观
景观面积：6 万 m^2

St-Moritz
圣莫丽斯

FEATURES 项目亮点

Along the lakeshore, tall trees and shrubs are arranged according to the topography and their seasons' changes to present different beauties all year round. Open and calm water as well as the reflections will remind people of the beautiful Swiss mountains and lakes.

岸边依地形变化广植高大乔木，湖岸栽植注意季相变化，水面宁静开阔，倒影依稀，四季不同，似有瑞士高山湖泊之美。

Overview

As the top-class residential project developed by Shenzhen Hwaloilee Group, St-Moritz is situated in a quiet valley of Meilin Mountain. The whole residential area is divided to three parts: zone A, B and C. Zone C in the north is designed for high-rise residential buildings. The landscape area here is about 60,000 m^2. With the driveways surrounding the community, the cars and pedestrians are totally separated. The large artificial lake and landscape courtyard covering an area of 3,000 m^2 are built entirely over ground. The site includes four terraces with an altitude difference of 10 m to present a well scattered and beautiful view.

项目概况

圣莫丽斯为深圳华来利集团旗下的顶级高尚住宅区，坐落在梅林山支脉的一个十分幽静的山谷中。整个住宅区分成 A、B、C 三个区。C 区位于用地北段的高层住宅区，景观设计面积约 6 万 m^2，小区四周环以车道，人车完全分流。3 000 m^2 大型人工湖和景观庭院完全建在地之上，场地分为四个台地，10 m 的落差使景观错落有致。

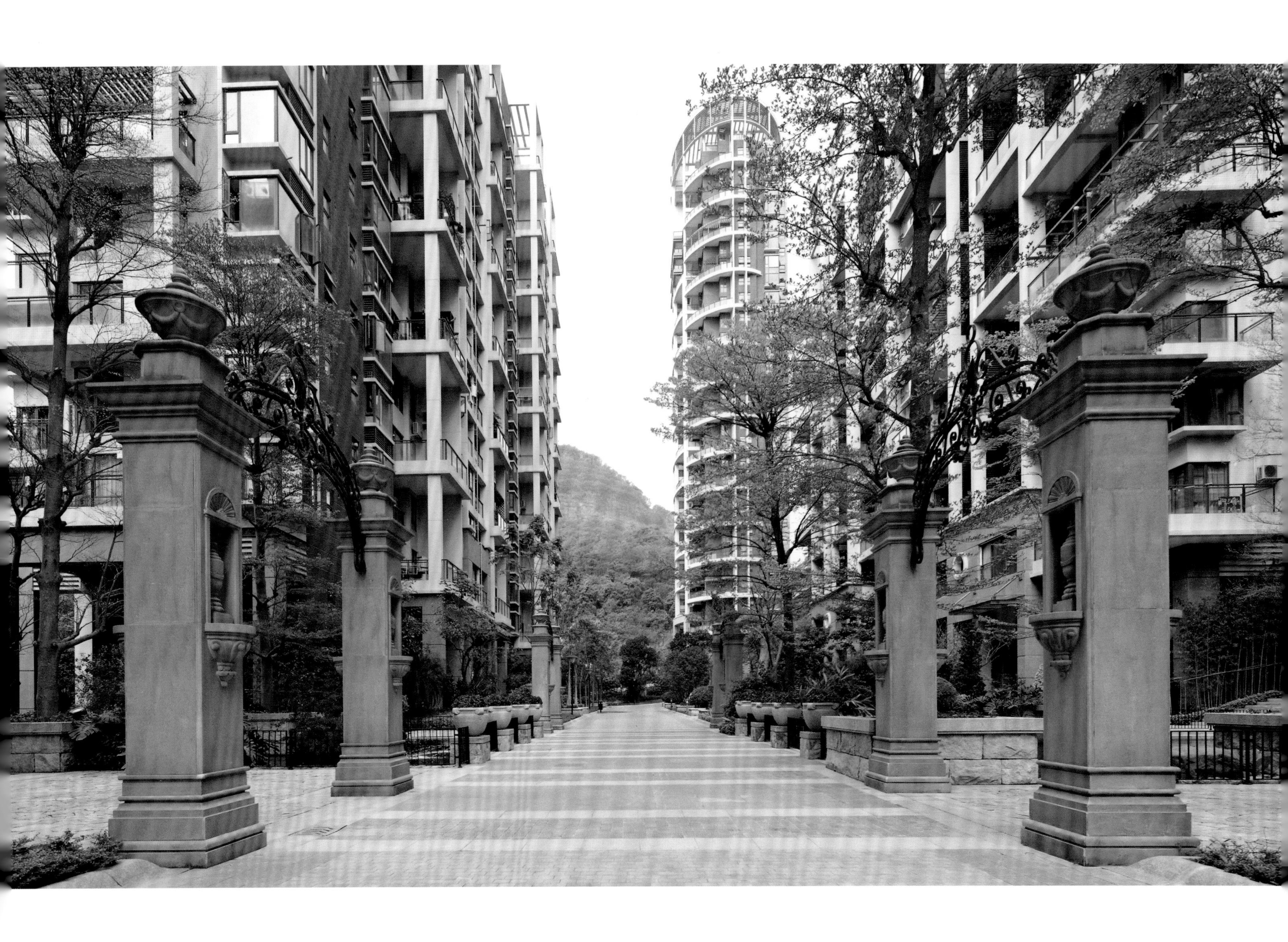

Site Plan 总平面图

Design Objective

The landscape is designed in Swiss mountainous town style. The landscape design for Zone C should not only shape its own identity but also keep harmonious with the landscapes in Zone A and Zone B.

设计目标

景观设计风格为瑞士高山小镇风情。C 区景观设计不仅要考虑拥有本区的个性特点，更要与建成的 A、B 区景观有机地结合起来，浑然一体。

Landscape Planning

By building the central lake area (Gao Xia Ping Hu) and the cascade streams (Die Ying Gua Bi), it connects the water system in Zone C with those rivers in Zone A and B, just like the source of rivers. The central lake area covers an area of about 3,000 m^2. It is situated in the middle of the terraces, surrounded by a ring road with a big altitude difference. The paths, terraces and plank roads are interwove to form a beautiful image and present varying landscapes. Along the lakeshore, tall trees and shrubs are arranged according to the topography and their seasons' changes to present different beauties all year round. Open and calm water as well as the reflections will remind people of the beautiful Swiss mountains and lakes. The music fountains are designed in the lake which will dance with music during weekends or holidays and shine beautifully at night. On the lakeside, there is a sculpture square with sculptures, pavilions, corridors and flower beds in classical Western style, just like a place of peace which is ideal for visiting and sightseeing. Water flows under the bridge and then reaches the cascade stream area to form the waterfalls, fountains or small ponds. Cyperus alternifolius, Thalia dealbata and water lilies are planted along the stream, and the rocks are stacked to form small hills. Several Syzygium jambos stand by the wooden plank road, and the creek runs merrily and finally meets in the rivers of Zone A and B. It continues to talk the story of St-Moritz with songs.

景观规划

通过修建中心湖区（高峡平湖）与跌落溪流（叠影挂壁）使C区水系与A、B区水系（小河）连接如一，似小河之源。中心湖区约3 000 m^2，设置在高台中央，外有环道围合，高差变化较大，通过小径、台地、栈道组织成交织画面，将高差融于变化的景观之中。岸边依地形变化广植高大乔木，湖岸栽植注意季相变化，水面宁静开阔，倒影依稀，四季不同，似有瑞士高山湖泊之美。湖中设有音乐喷泉，节假日随音起舞，夜间灯光璀璨，美轮美奂。湖滨为雕塑广场、雕塑、亭廊、花池，取西式古典为主题，或观景或游览，仿若世外城邦。湖水流过小桥，进入叠落溪流景观区，或瀑布或喷泉或小潭，依坡顺流而下，光影动感交错。溪边植风车草、再力花、睡莲，山石叠砌，浑似天然。又植水蒲桃数棵于木栈道边，山溪欢快流淌，最后进入地下流入A、B区的小河中，汇为一体，它如欢歌再续着圣莫丽斯动人的故事。

KEYWORDS 关键词

French Romance
法式浪漫

Soft Landscape
软质景观

Urban Space
城市空间

European Style
欧式风格

Location: Lianyungang, Jiangsu
Landscape Design: Baroque Design Group
Land Area: 120,000 m^2

项目地点：江苏省连云港市
景观设计：香港博唯规划建筑与环境景观设计有限公司
占地面积：12 万 m^2

Excellence · Cote Dazur
卓越 · 蔚蓝海岸

FEATURES 项目亮点

With the concept of creating an urban space with French atmosphere, designers adopt lots of romantic flowers and grass as the main soft landscape to construct a sweet home that is luxury, warm and touched.

景观设计重在打造法式浪漫的城市空间。多处采用浪漫的花草为主要软质景观，营造一个华丽而温暖，又沁入人心的温馨家园。

Overview

Nestled in the center of Coast New Town, this development is at the west of Huaguoshan Road and north of East Avenue that has connected the east and west of Lianyungang City. It is also close to Municipal Administration for Industry and Commerce, Command Center of 110, Administration Committee of the Development Zone Building, National China-Japan (Lianyungang) Ecological and Industrial Park, National Software Park, Lianyungang Pharmaceutical Industrial Research and Development Center, as well as 140,000 m^2 Polaris Plaza, occupying advantaged location and having an opportunity for the appreciation.

项目概况

连云港卓越·蔚蓝海岸位于海滨新城核心，东靠花果山路、南接东方大道，连接贯穿城市东西的主干道，交通顺畅。项目毗邻市工商局、市 110 指挥中心、开发区管委会大厦、国家级中日（连云港）生态科技产业园、国家级软件园、连支港医药产业研发中心、14 万 m^2 北极星商业广场，区域条件优越，升值空间巨大。

Design Objective

Following the concept of "establishing the art of value", designers integrate the idea of quality, gardens, courtyards and details into the planning, design and marketing of the project, using new high-end French villas to establish the core value, and deepening to build natural and healthy residences with French life style, so that residents are able to enjoy the harmony between architecture and nature, as well as the sense of comfort, honor and dignity.

设计目标

项目秉承 “构建价值的艺术” 的理念，将“品质、园林、庭院、细节”产品理念贯穿于规划、设计、营销之中，项目以新法式高端别墅形象切入塑造产品核心价值，通过产品核心价值营造真正健康自然的居所，引领法式国际居住生活模式，让客户享受建筑与自然的和谐，充分体现居住舒适感、荣誉感和地位感。

Landscape Planning

With the concept of creating an urban space with French atmosphere, designers adopt lots of romantic flowers and grass as the main soft landscape to construct a sweet home that is luxury, warm and touched. They are loyal to the design concept of people orientation, diversity of landscape, as well as reasonable functional design and design criteria. With the integration of classic and modern, entirety and detail, coexisting of beauty and sensibility as well, it has represented the most exhaustive romance to residences by the French landscape that collects the functions of leisure, living and recreation.

景观规划

景观设计重在打造法式浪漫的城市空间。多处采用浪漫的花草为主要软质景观，营造一个华丽而温暖，又沁入人心的温馨家园。设计忠于以人为本的设计思想、功能设计的合理性、景观设计的多样性、设计尺度的合理性。古典与现代交替，整体与细节融合，唯美与感性共存，集悠闲、居住、娱乐为一体的法式园林景观呈现给人们一种最彻底的浪漫。

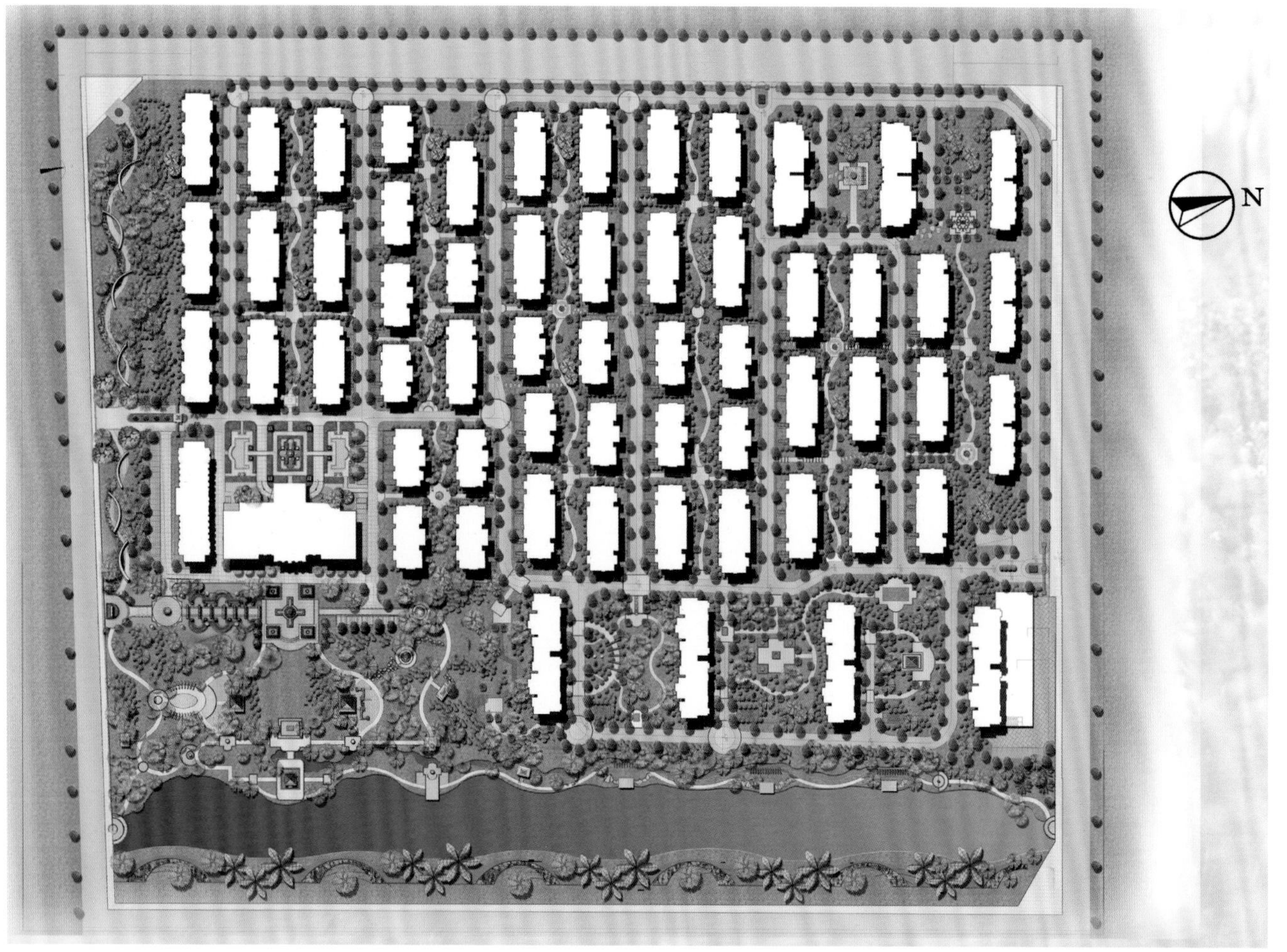

Site Plan 总平面图

KEYWORDS 关键词

Deep Space
纵深空间

Water System Division
水体分割

Feature Planting
特色植栽

European Style
欧式风格

Location: Nanhu New District, Jiaxing, Zhejiang
Landscape Design: Botao Landscape (Australia)
Land Area: 400,000 m^2

项目地点：浙江省嘉兴市南湖新区
景观设计：澳大利亚・柏涛景观
占地面积：40 万 m^2

Eastern Provence
东方普罗旺斯

FEATURES 项目亮点

The water system in villa area became the natural boundaries of dividing other villas. Designer tries to use humanistic art pieces to create elaborate details for villa area's landscape.

别墅区的水体形成了分割其他别墅的自然界限，设计师力图利用有人文情节的艺术小品为别墅区的景观创造精巧的细节。

Overview

The project is located in Nanhu New District, Jiaxing City, 5 km distance to the south of city center, and needs to drive only 15 minutes to the downtown.

项目概况

项目地处嘉兴市南湖新区，位于市中心南向 5 km 处，从项目用地到达中心城区仅 15 分钟车程。

东方·普罗旺斯
东方·普罗旺斯

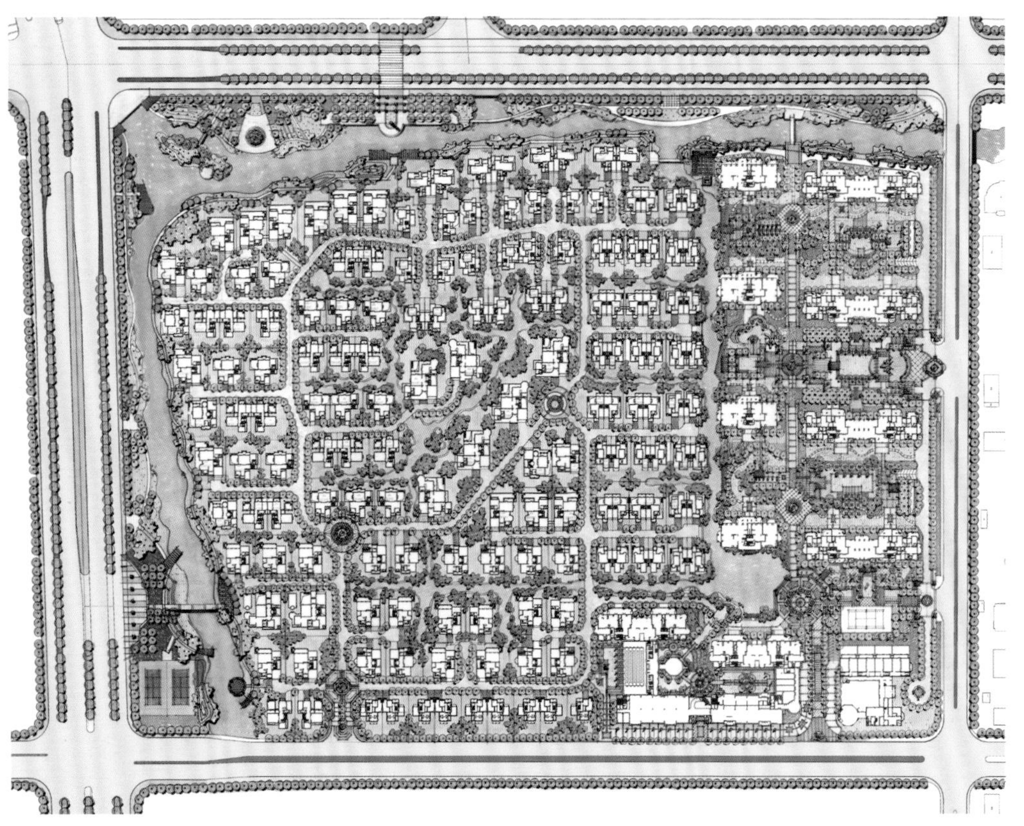
Site Plan 总平面图

Design Objective

The project is a continuation of the Mediterranean style, and its planning and design intend to create classic luxury house with artistic conception of French Provence through previous product positioning. The history of time sharpens into magnificent humanistic views of multi-architectural styles and multi-cultural monuments, and also leaves a colorful past on Provence. Time passes, Provence has perfectly blended ancient and modern fashion, thereby deposits into a haunting paradise.

设计目标

项目是地中海风格的延续，策划和设计经过前期的产品定位，意在打造法式普罗旺斯意境的经典豪宅。历史上时光的磨砺给普罗旺斯留下了一个多建筑风格、多文化遗迹的瑰丽人文美景，同时也赋予普罗旺斯一段多姿多彩的过去。岁月流逝，普罗旺斯将古今风尚完美地融合在一起，从而沉积出一个让人流连忘返的人间乐土。

Landscape Planning

High rise zone—site of high-rise block, has larger spatial span and greater deep space. The population density of high rise zone is larger than that of villa area, requiring wider public venues. Therefore, landscape designers created three different theme squares which respectively center as western fountains, sharp square monument, and a large column pavilion. The site on the north side is relatively narrow, continuing from west to east. The commercial podium in northwest corner and the site edge of high-rise apartments are relatively uneven, so designer designed the landscape reflected the theme of natural beauty as landscape green transition zone that connects high level. There are landscape elements such as natural water system, close planting trees, twisting waterfront and raised hills. From here, people began to enter the landscape veranda zone where landscape is divided into four units based on small and middle landscape elements such as gazebo, corridor, sculptures, small fountains, flower pots and others, with a ribbon-like landscape corridor.

Villa Area—the characteristics of the product determines the site's private temperament in relatively narrow villa area. Landscape designer here themed as courtyard, takes the small space landscape elements as the core such as flower pots, private revetment, elegant walls, lights, scene sculpture and feature planting, creating courtyard space in different theme. Here, designer puts more emphasis on exclusive venues, reducing the use of public space. The water system in villa area becomes the natural boundaries of dividing other villas. Designer tries to use humanistic art pieces to create elaborate details for villa area's landscape.

景观规划

高层区——高层地块的场地，空间跨度较大，有更大的纵深空间，高层区的人口密度大于别墅区的人口密度，需要更宽阔的公共活动场地。因此，景观设计师创作了三个不同主题的广场，分别以西洋喷水池、尖方碑和大型柱亭为核心。北侧的场地较狭长，由西向东延续发展。西北角的商业裙楼与高层公寓的场地边缘较参差，设计师设计了体现自然风光主题的景观，作为连接高层的景观绿色过渡区。这里有景观元素的自然水体、密植的树林、波折的水岸、隆起的丘陵等。从这里人们就开始进入景观游廊的地带。这里的景观以凉亭、连廊、雕塑、小喷泉、花钵等中小体量的景观元素为核心，分为四个单元，还有一条带状递进的景观走廊。

别墅区——在相对狭窄的别墅区，产品的特性决定了场地的私有气质。景观设计师在这里以庭院为主题，以小空间的景观元素，如花钵、私家驳岸、优雅的围墙、灯具、情景雕塑、特色植栽为核心，创造不同主题风格的庭院景观空间。在这里，设计师更强调的是场地独享，降低公共空间的使用度，别墅区的水体形成了分割其他别墅的自然界限，设计师力图利用有人文情节的艺术小品为别墅区的景观创造精巧的细节。

KEYWORDS 关键词

Artistic Beauty
艺术美感

British Style
英伦风格

Chinese-Western Mixed Style
中西合璧

European Style
欧式风格

Location: Xinxiang, Henan
Developer: Xinxiang Lvdu Properties Co., Ltd.
Landscape Design: Botao Landscape (Australia)
Landscape Area: 90,000 m^2

项目地点：河南省新乡市
开发商：新乡市绿都置业有限公司
景观设计：澳大利亚・柏涛景观
景观面积：9 万 m^2

Xinxiang Lvdu · Windsor Castle

新乡绿都・温莎城堡

FEATURES 项目亮点

The iron decorations, sculptures, dormers and slope roof are well interpreting the solemnity, simplicity and elegance of the British-style architecture.

铁艺、雕塑、老虎窗、坡屋顶都充分诠释英伦建筑的庄重、古朴和典雅。

Overview

As the only pure British-style villa project in Xinxiang City, Windsor Castle provides a collection of superposed villas ranging from 210 m^2 to 530 m^2. The floor plans are varied and flexible to meet different requirements. The iron decorations, sculptures, dormers and slope roof are well interpreting the dignity and elegance of the British-style architecture.

项目概况

新乡绿都・温莎城堡是新乡唯一的纯英伦别墅社区，典藏 210 ~ 530 m^2 叠加别墅。户型灵活多样，空间灵活使用，打造新乡南区标杆住宅。铁艺、雕塑、老虎窗、坡屋顶都充分诠释英伦建筑的庄重、古朴和典雅。

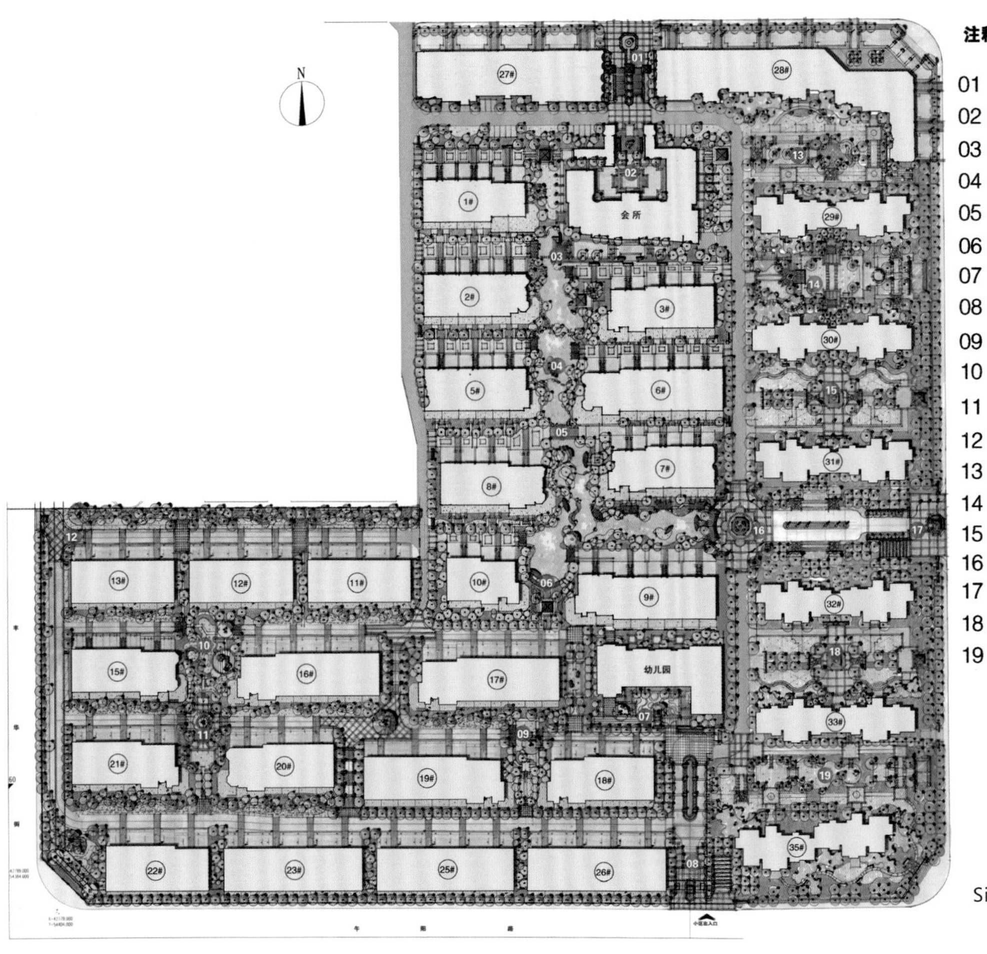

注释:

01 北侧主入口广场
02 会所中庭
03 景观平台
04 景观跌溪
05 景观桥
06 观景平台
07 幼儿园活动场
08 南侧主入口广场
09 宅间草坪庭院
10 宅间儿童活动广场
11 树荫广场
12 西侧车行出入口
13 清林小径庭院
14 赏心庭院
15 品塑庭院
16 中心水景广场
17 东侧主入口广场
18 休闲庭院
19 畅心庭院

Site Plan 总平面图

Design Objective

The landscape design aims to create an elegant, beautiful and functional British-style garden which allows people to have different life experiences. Inspired by the prestigious British-style houses and villas with charming and tranquil gardens, the development will bring the neighborhood with great modern convenience. It will be a British-style community integrating both the traditional cultures and the modern Chinese lifestyle.

设计目标

此次项目景观设计理念和景象将会展示出一个优雅、美观大方和功能性的英式景观花园，一个可以让人们拥有不同生活体验的地方。设计师深受著名的拥有吸引人的宁静花园的英式豪宅和小别墅的启发，因此该项目将会为所有居住在附近的居民提供现代化的方便。这将会是一个令传统文化和当代中国生活融入英式时尚的社区。

Landscape Planning

Composed of various types of residential facilities such as the high-rise residential towers and low-rise townhouses, this well-planned development will provide ample amount of functional, recreational and aesthetically beautiful outdoor and landscape facilities. Furthermore, it aims to achieve a kind of British country atmosphere.

The project is strategically divided into specific areas/zones to enhance the British-style atmosphere and the country style. Namely,
Gateway squares: This is the main entrance of the development. It has a very grand and welcoming atmosphere that is unforgettable for everyone.
Gateway Shops: The Gateway Shops is the zone for commerce and retails which will provide the residents with British-style shopping experience.
South Gate: A secondary entry in the southern residential area.
East Gate: The East Gate provides another magnificent zone where the high-rise towers can be built.
Oasis squares: It is simply named because of the passive function of this area. Here the waterscape garden provides some facilities for relaxation and activities.
Lakeside Manor: It is a residential area near to the "Oasis Plaza", which provides a more leisurely and pastoral waterscape garden.
Garden Estate: The townhouses are built around the pond. In the center, there is a multi-functional garden with complete facilities for the old and the children.
Mansion Gardens: Mansion Gardens is an area for high-rise residences and courtyard gardens. As a highly populated area, it needs more green and public areas.
The Chateau: It refers to the clubhouse and its outdoor courtyard area.

景观规划

项目规划合理，由包括高层居住塔楼和低层联排别墅在内的多种住宅产品组成，提供了能满足人们功能性、娱乐性和艺术美感需求的户外和景观设施。更进一步说，项目旨在打造一种英国乡村氛围。

项目被合理划分为不同的功能区域来强化其英式田园氛围。这些区域包括：
入口广场：作为该住宅小区的主要入口，这里洋溢着热情而友好的氛围，令人印象深刻。
入口商铺：入口商铺区是用于商业和零售的区域。居民将能在此体会到英伦风情的购物体验。
南门：南门是位于南住宅区的次入口。
东门：沿东门布置了另外一块高层住宅区。
绿洲广场：之所以被这么简单地命名，是因为这片区域更倾向于一种被动的功能。这里的水景花园为人们提供一些休闲活动的设施。
湖畔庄园：湖畔庄园是临近绿洲广场的居住区。这里建有一个更加休闲、更富有田园气息的水景花园。
花园别墅：联排别墅围绕水塘而建，中间的多功能花园设施齐全，是老少皆宜的休闲场所。
大厦花园：大厦花园分布着高层住宅和庭院花园。由于居住的人口相对密集，因此需要配备更多的绿化和公共区域。
城堡会所：包含会所及其户外庭院区。

KEYWORDS 关键词

Private Landscape
私密景观

Central Water Feature
中心水景

Welsh Flavor
威尔士风光

European Style
欧式风格

Location: Changsha, Hunan
Landscape Design: Open Fields (Beijing) Landscape Design Co., Ltd.
Land Area: 50,000 m²

项目地点：湖南省长沙市
景观设计：北京易德地景景观设计有限公司
占地面积：5 万 m²

Welsh Spring
威尔士春天

FEATURES 项目亮点

Designers put nature first and place high-rise residences and enclosed gardens in elliptic to avoid monotonousness, boredom and lack of privacy that will be easily caused by normal determinant buildings.

在规划设计上以自然为先，打破一般行列式建筑造成的单调、对视与枯燥感，巧妙将高层与围合式花园成椭圆型排布。

Overview

Located in the most suitable living plot of Changsha, Welsh Spring enjoys three garden groups as Yuehu Park, Yuedao Park and landscape belt of Liuyang River, moreover, Changsha nursery garden and Malanshan are also nearby. With this advantaged location but also close to urban area, it connects with each trading area of Changsha through Sany Avenue and Wanjiali Road, 10 minutes to train station and 15 minutes to Wuyi Plaza.

项目概况

威尔士春天位于长沙最宜居的月湖板块。月湖公园、月岛公园、浏阳河风光带三大公园群，以及市苗圃，马栏山环绕。如此世外桃源，却不藏于郊野。经三一大道、万家丽路通达长沙各大商圈，10 分钟可到火车站，15 分钟可到五一广场。

Design Objective

Inspired by natural landscape of Welsh, England, the development has the green coverage ratio of 64% with romantic and England style ecological gardens, to create private landscape spaces for residents. Also, the 2,000 m^2 central water feature reveals unique fascination of nature.

设计目标

威尔士春天源自于英格兰威尔士自然景观创作灵感，64% 绿化率，浪漫而富有英伦格调的生态园林，为居者打造“人在园中，藏在景中”的私密观赏空间；2 000 m^2 中心水景，在潺潺流水、鸟语蝉鸣中，体会不一样的自然魅力。

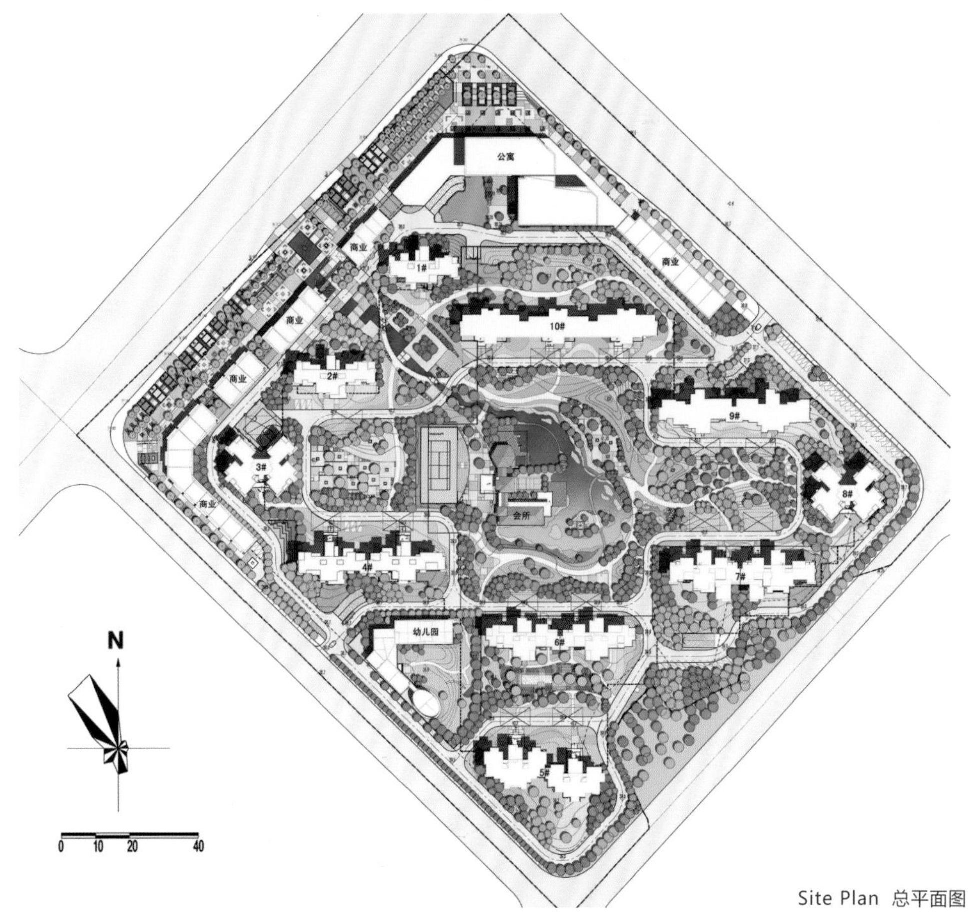

Site Plan 总平面图

Landscape Planning

In order to avoid monotonousness, boredom and lack of privacy that will be easily caused by normal determinant buildings, designers put nature first, placing high-rise residences and enclosed gardens in elliptic type, cooperating with reasonable spatial layout, keeping practicability and functionality in mind to ensure fully use of outdoor landscape. For instance, landscape environment of green coverage rate reaching 64%, super-wide building interval reaching 130 m and 30,000 m^2 central England garden.

景观规划

在规划设计上以自然为先，打破一般行列式建筑造成的单调、对视与枯燥感，巧妙将高层与围合式花园成椭圆型排布，通过空间合理布局，以实用性、功能性为主导，最大尺度保证建筑对外部景观的利用。如 64% 超高绿化率、130 m 超宽楼间距、3 万 m^2 中心英伦园林等景观环境的打造。

Southeast Asian Style

东南亚风格

Tropical Atmosphere
热带风情

True Nature
返璞归真

Feature Waterscape
特色水景

KEYWORDS 关键词

Harbor Culture
港湾文化

Local Flavor
本土气息

Modern Sense
强烈现代感

Southeast Asian Style
东南亚风格

Location: Haizhu District, Guangzhou, Guangdong
Landscape Design: Palm Landscape Architecture Co., Ltd.
Land Area: 18,483 m^2

项目地点：广东省广州市海珠区
景观设计：棕榈设计有限公司
占地面积：18 483 m^2

World Coast
君华天汇

FEATURES 项目亮点

Plenty of modern lines are adopted for the development, each landscape node is influenced and integrated into a whole through the connection of landscape clues. Landscape designers provide various kinds of spatial experience to residents through careful treatment on terrain height differences.

在项目中，充满强烈现代感的线条来塑造小区空间，各个主要景观节点通过景观线索，相互影响、渗透成为一个有机的整体。通过对地形高差的处理，在小区形成丰富的空间体验。

Overview

World Coast will provide two commercial and residential buildings, while one is 18-storey high-rise residential building and another is 30-storey luxury units for sales with view of Pearl River. It enjoys sophisticated transportation network, with Renmin Bridge and Jiefang Bridge connecting the two banks, and Binjiang West Road leading to inner ring road makes it acceptable to the whole city. The perfect public traffic system has also been completed with a lot of bus lines such as No. 10, No. 270 and 248 as well as completion of Metro Line 6. It is located between two CBDs of Bai'etan and Zhujaing New Town, while with financial street exactly setting opposite, it occupies the core resources of three CBDs.

项目概况

君华天汇将建成 2 栋商住一体的住宅楼，包括一栋 18 层住宅楼以及一栋销售的 30 层望江豪宅项目。项目交通路网完善，人民桥、解放桥等跨江大桥贯通南北，滨江西路连接内环线通达全市。公交系统的完善也为出行带来了便利，10、270、248 等多路公交车通往市内各区。地铁 6 号线的贯通，也将为项目提速。地处白鹅潭 CBD 与珠江新城 CBD 之间，江对岸是金融街，掌控 3 大 CBD 核心资源。

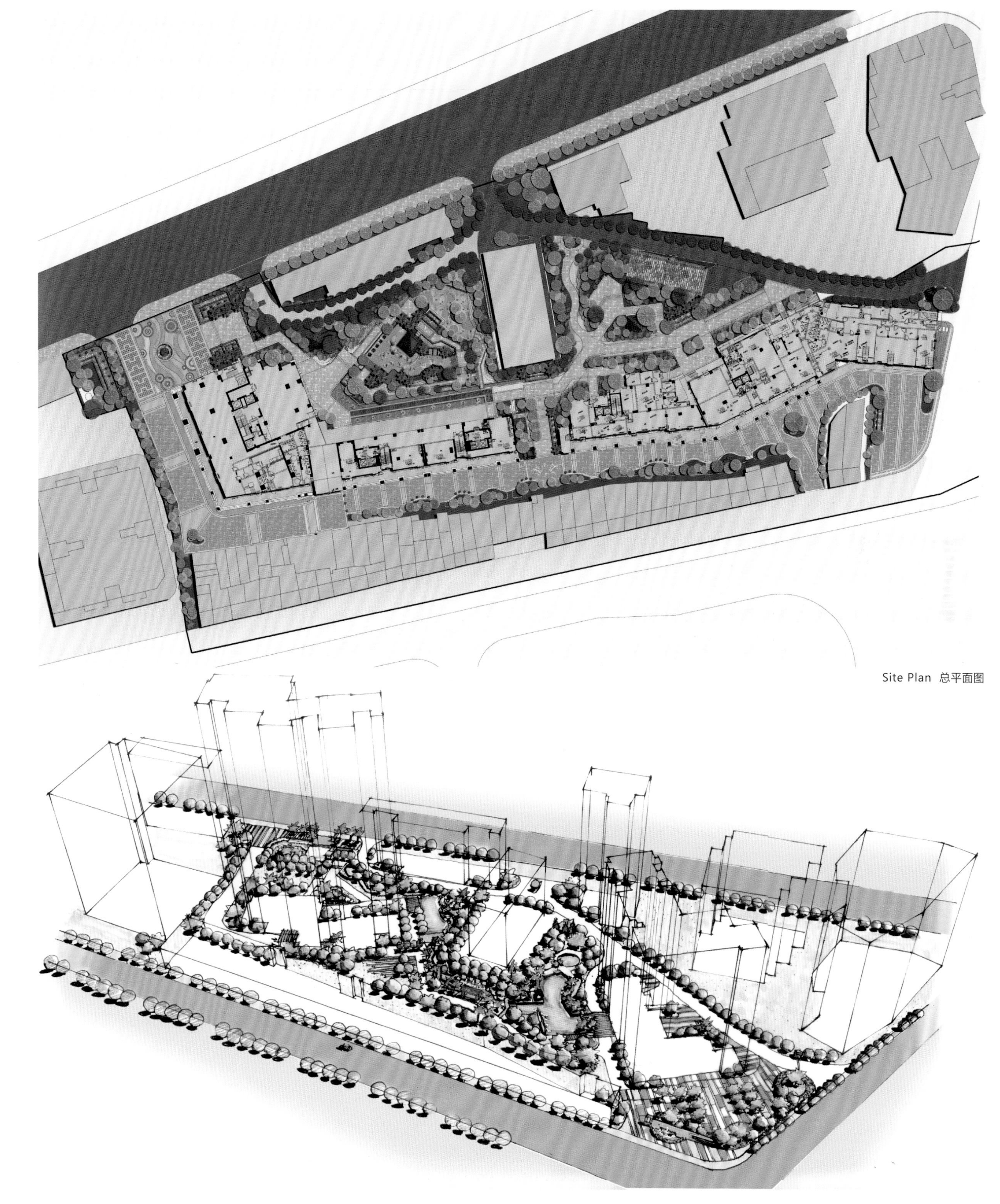

Site Plan 总平面图

君华·天汇
WORLD COAST
湾区生活体验馆

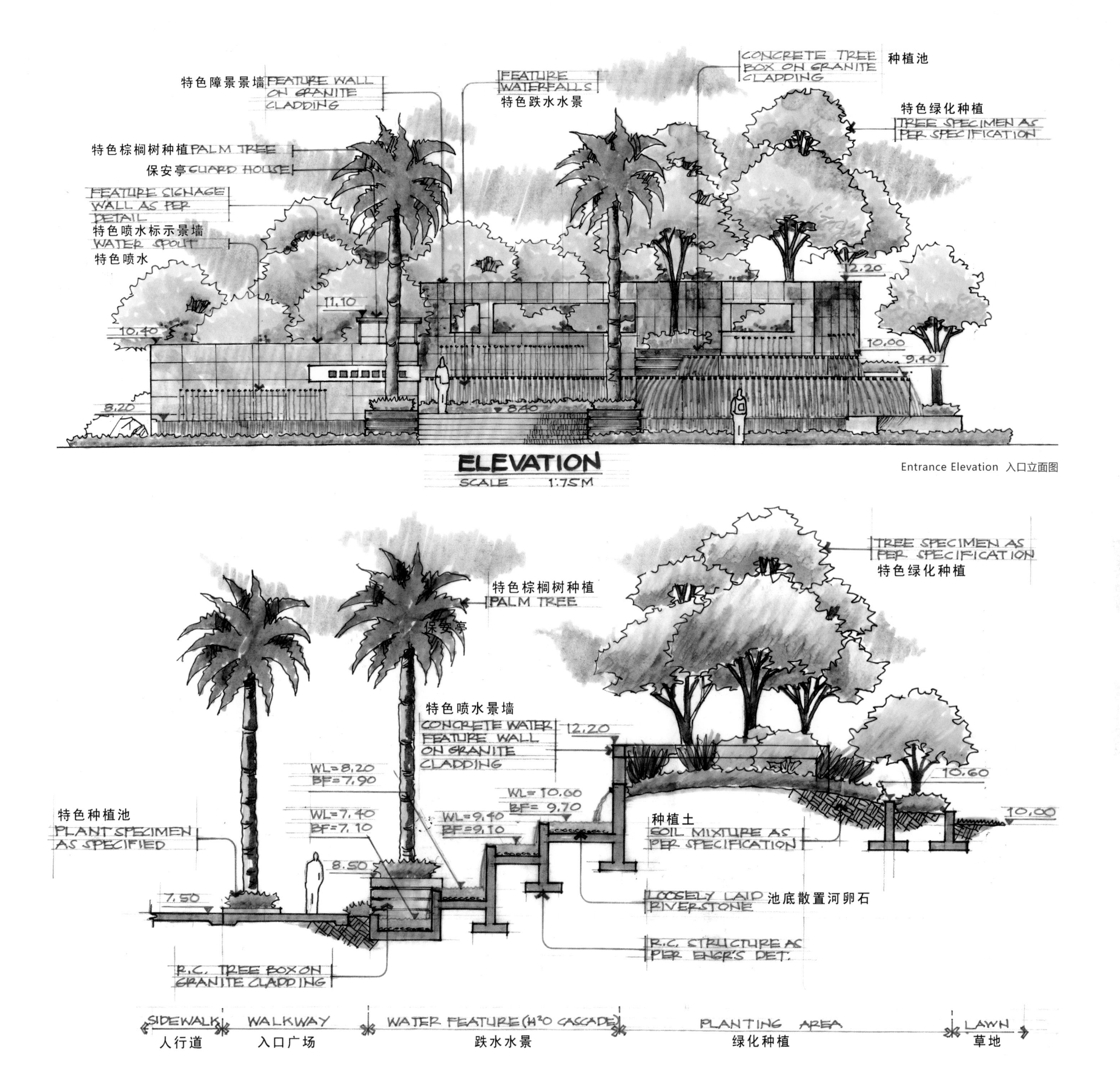

Entrance Elevation 入口立面图

Main Entrance Section 主入口剖面图

Design Objective

Landscape designers are trying to take full advantage of location to introduce the gorgeous riverview with local features into the development through complex landscape methods, to create a modern, natural and pleasant residential community but also with local flavor. The project extends the beauty of nature and ecology to represent unique comfortable flavor of Southeast Asia, cooperating with construction of landscape, guidance of views and sightseeing fun of varied landscape step by step to inspire visitors' sensation, also expressing harbor culture connotation through accessories and details to meet the higher psychological needs regarding living environment.

设计目标

设计师试图凭借良好的地理优势，通过视线的引导和复合式景观手法的处理，将充满广州本土气息的一道亮丽江景引入小区之中。设计师力求将小区打造成一个具有强烈现代感，自然舒适又富含本土气息的生活社区。项目延续自然，传承一种生态美，体现现代东南亚风情的假日舒适情节，通过景观的营造、视线的引导、步移景异的游览乐趣，充分调动观赏者的五官感受，从小品与景观细节体现港湾文化内涵，激发联想，满足人们对于环境的更高层次的心理需求。

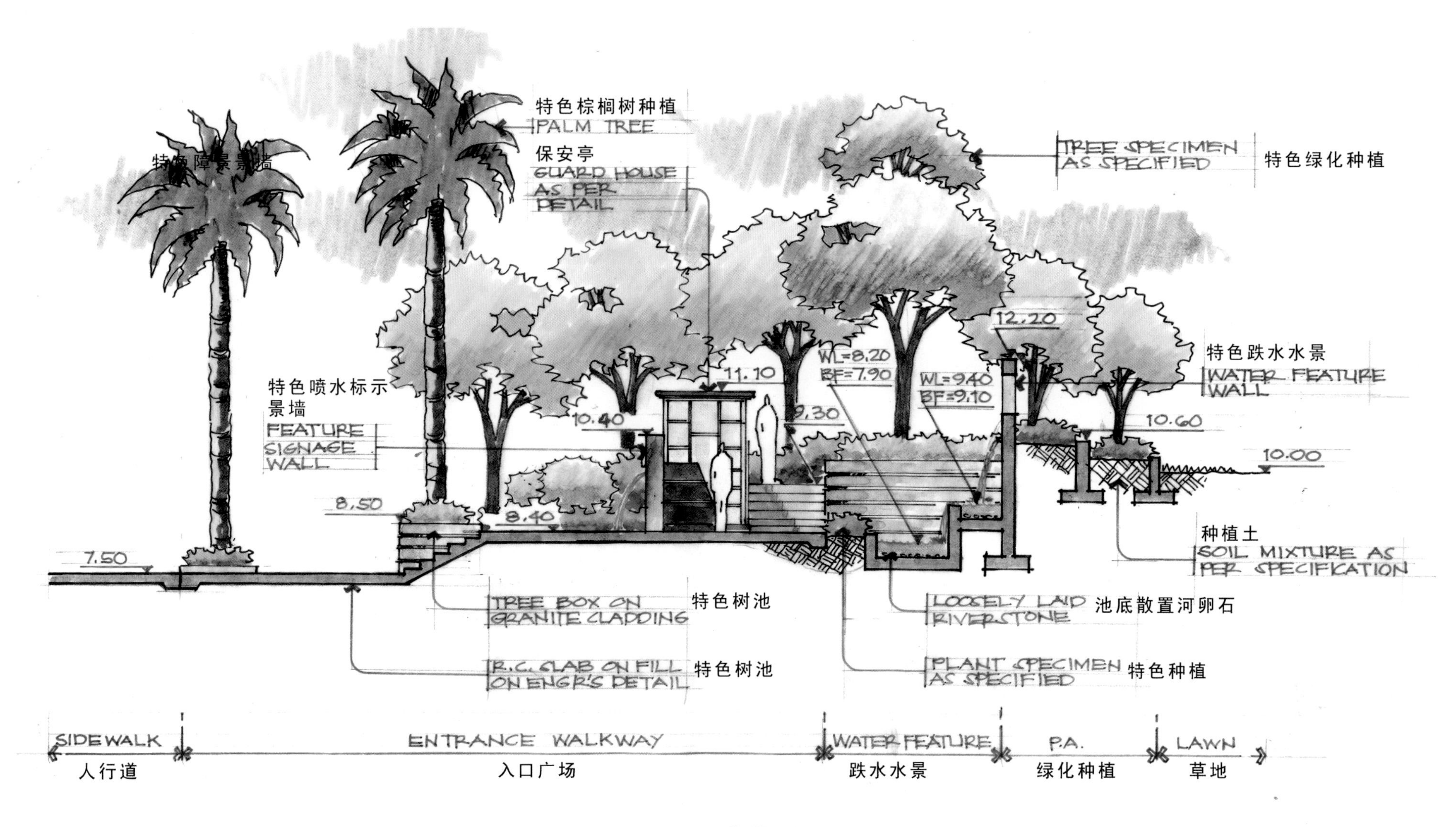

Entrance Section 入口剖面图

Leisure Landscape Section 休闲景观剖面图

Landscape Planning

Plenty of modern lines are adopted for the development, each landscape node is influenced and integrated into a whole through the connection of landscape clues. Landscape designers provide various kinds of spatial experience to residents through careful treatment on terrain height differences, but also building reasonable division between development and original buildings of the site. Different shapes of water features are placed all over the development to be regarded as the soul of the whole landscape system to create leisure and pleasant landscape. Landscape designers carefully design each landscape theme to further represent the design concept to improve the visual and sensory experience of landscape dramatically.

景观规划

在项目中，充满强烈现代感的线条来塑造小区空间，各个主要景观节点通过景观线索，相互影响、渗透成为一个有机的整体。通过对地形高差的处理，在小区形成丰富的空间体验。同时起到将小区与原有保留建筑形成合理的分隔。水系以不同的姿态贯穿全园，成为整个景观的灵魂，塑造悠然的山水景观。项目通过对景观分区不同的主题提炼，进一步来诠释概念，使得小区景观在视觉上和感受上都得到了艺术的升华。

KEYWORDS 关键词

Subtropical Flavor
亚热带风情

Curves Layout
曲线构图

Water Features Landscape
水系景观

Southeast Asian Style
东南亚风格

Location: Baiyun District, Guangzhou, Guangdong
Landscape Design: Guangzhou Homy Landscape Co., Ltd.
Land Area: 33,000 m^2

项目地点：广东省广州市白云区
景观设计：华誉国际景观设计公司
占地面积：3.3 万 m^2

Times Peanut II
时代花生二期

FEATURES 项目亮点

Emphasizing on the expression of subtropical garden landscape and joining with modern elements, designers adopt smooth and concise lines to create landscape spaces with clear functions division.

项目着重于表现亚热带风情园林景观，同时加入现代元素，以流畅的线条和简练的直线构图，组成功能明确的景观空间。

Overview

Times Peanut II is located in Tonghe, Baiyun District which is belonging to Guangzhou North Avenue trading area, and connecting with city expresses such as Huanan Express and Shatai Road, occupying convenient transportation with 4 stops of Line 3 away from Tianhe District, only 15 minutes' drive to Tianhe CBD and more than 40 bus routes. It also has perfect supporting facilities around including four shopping malls, Jiayu Sunny Plaza, Southern Hospital, Tonghe Primary School and so on.

项目概况

时代花生二期位于白云区同和，属于广州大道北商圈，与华南快速、沙太路等城市干线快速连线，地铁 3 号线 4 站即达天河，开车 15 分钟直达天河 CBD，40 多条公交线路，交通便利。项目周边配套成熟，囊括四大购物中心、嘉裕太阳城广场、南方医院、同和小学等，生活便利。

Design Objective

Emphasizing on the expression of subtropical garden landscape and joining with modern elements, designers adopt smooth and concise lines to create landscape spaces with clear functions division, making the whole space reasoningly but also full of passion.

设计目标

项目着重于表现亚热带风情园林景观，同时加入现代元素，以流畅的线条和简练的直线构图，组成功能明确的景观空间，使整体布局赋予理性，又饱含激情。

Landscape Planning

The transmeridional water features are employed as landscape axis to unify the landscape design for the whole community. For the design of central landscape space, designers ingeniously use the contrast of movement and quietude, openness and privacy, as well as modern and nature to make spaces full with rhythm and rhyme. In the meanwhile, a few small-sized water features are used in the elevated level so that the space would be much more dynamic and colorful. Smooth, vivid and free curves are selected for fire engine access, also coupling with grass to let roads in the community become interesting and also in various shapes, so that residents are able to embrace and enjoy the comfortable and modern spaces.

景观规划

贯通东西向的水系景观作为景观轴，统一整个小区的设计。在中心景观空间的设计上，运用了动与静、开放与私密、现代与野趣的对比，创造出有节奏、有韵律的空间。同时在架空层设计了多个小型水景，使空间变得活跃和丰富。消防通道的设计采用流畅、活泼、自由的曲线构图，并与草坪绿化相结合，使小区道路变得丰富而有趣，居民可尽情地拥有和享受这个洋溢着现代气息的舒适空间。

KEYWORDS 关键词

Green Space System
绿地系统

Courtyard Green Land
庭院绿地

Layout of Mountain and Sea
山海格局

Southeast Asian Style
东南亚风格

Location: Lingshui County, Hainan
Developer: Guangzhou R&F Properties Co., Ltd.
Landscape Design: Guangzhou Kemei Urban Landscape Design and Planning Co., Ltd., Hainan Branch
Design Team: Huang Junjiang, Zhang Xiaobin, Wang Youcheng, Liao Haina, Zhao Yinfang, Deng Xiaokai
Land Area: 3,200 mu (approx. 2,133,333.4 m^2)
Completion: 2014

项目地点：海南省陵水县
开发商：富力地产
景观设计：广州市科美都市景观规划有限公司海南分公司
设计团队：黄俊江、张小兵、王佑程、廖海娜、赵银芳、邓小凯
占地面积：3 200 亩（约 2 133 333.4 m^2）
建成时间：2014 年

Hainan Future Villa

海南富力湾

FEATURES 项目亮点

Courtyard gardens are carefully designed in the residential groups, selecting exquisite pergolas, artificial mountains, sculptures, accessories, flower beds and water features to highlight uniqueness of entrance.

在组团内部，建筑周围设置细腻、精致的庭院园林，运用花架、假山、雕塑、小品、花坛、流水，强调组团入口特色。

Overview

Situated in the southern end of Zone B of Xiangshui Bay Tourist Area, Lingshui County, Future Villa is 18 degree north latitude which is the connection area of subtropics and tropics, lying on Diaoluo Mountain National Forest Park and close to Nanwan monkey island, national conversation area of macaque, in the south. It is planned to be developed by three stages to be a world-class tropical coastal resort area with a land area of 3,200 mu and green coverage ratio up to 64.8%. It is surrounded by mountains and sea while the west is wider and higher than the east, a south-north 4.2 km coastline is enjoyed by the project.

项目概况

富力湾位于陵水县香水湾旅游区 B 区南端，地处热带与亚热带交汇处，北纬 18° ，背依吊罗山国家森林公园，南临国家猕猴保护区——南湾猴岛。富力湾总体规划 3 200 亩，超大手笔打造国际一流热带滨海休闲度假社区，分三期开发，绿化率高达 64.8%。项目两面青山环绕，依山傍海，西高东低，面宽开阔，南北为长达 4.2 km 的私属海岸线。

一线海景洋房示范区
悦
阁

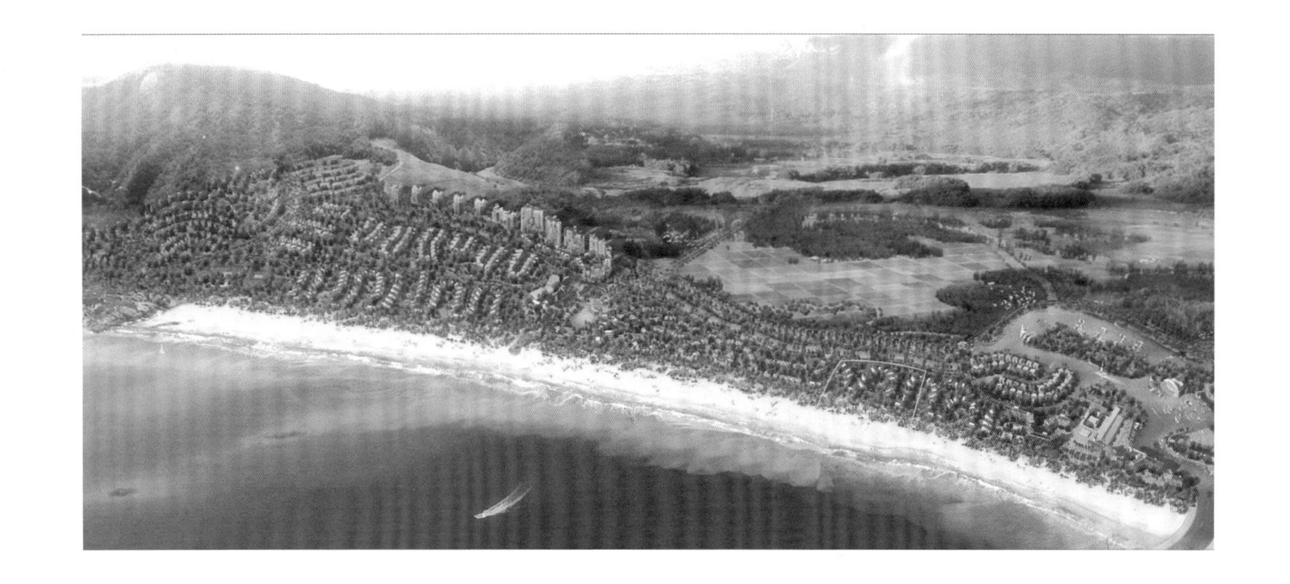

Design Objective

Future Villa is taking full advantage of natural resources and highlighting the original relationship between mountain and sea, insisting on sustainable development and respecting existing ecological environment to build a sophisticated landscape environment that integrated artificial and natural elements together perfectly. Green space system is comprised by park green land, green buffer, community and courtyard green land, while each green space is connecting with each other to form a complete green space system.

设计目标

项目充分利用天然资源的优势，突出山与海的格局特点，坚持生态环境优先，可持续发展原则，尊重现有的自然景观环境，建立完整的人工与自然融为一体的景观环境。景观绿地系统由公园绿地、防护绿地、社区环境绿地、庭院绿地组成。绿地间相互连同与渗透，形成点线面结合的完整绿地景观系统。

Landscape Planning

There are five landscape groups in total for the project, which are Zone M (coastal residences), Zone N/C (rainforest villa groups), Zone D/K/O (sand beach villa group and the headmost costal villas), Zone Q (duplexes) and Zone H (seascape apartment), as well as commercial area of yacht.

Artificial rainforest valley is placed at the main entrance, while driving in the valley with rainforest overhead, which will be welcoming and express strong sense of belongs, furthermore, the main entrance will enjoy recognizable and clear landscape. The tropical plants have blocked the far ocean view to provide different mental feelings while driving.

The flood plain of northwest of Gangpo River is approx. 2 m height, enjoys rare wetland landscape while it is planned to be an ecological wetland park, reserving the original topography and existing plants to be a natural habitat for birds. As the island will be inundated in flood period, there are no fixed tour facilities for visit, only tour paths and boat stops are set for visiting.

Courtyard gardens are carefully designed in the residential groups, selecting exquisite pergolas, artificial mountains, sculptures, accessories, flower beds and water features to highlight uniqueness of entrance. Features of landscape are echoing with that of architectural style, making it possible for each architecture to enjoy optimal landscape view and every residence to live in quiet and pleasant environment.

Calvin Klein
CK

BMW roadster

景观规划

景观组团由五部分组成：M 区海景住宅、N/C 区雨林别墅组团、D/K/O 区沙滩别墅组团及一线海景别墅、Q 区联排揽海别墅及 H 区海景公寓、国际游艇商业区。

在规划区入口处设置人工热带雨林山谷，车道在山谷中穿过，两旁茂密的雨林在高空连接在一起，行使其中，将产生强烈的到达感和仪式感，使主入口景观显著而富有特色。在视线上，蜿蜒的车型路利用热带的植物群对远处的海景形成了遮挡，使人们在不同的空间中有着不同的心理感受。

港坡河西北岸的河漫滩地高程 2 m 左右，拥有难得的入海口湿地景观；将其规划为生态湿地景观公园，保持原地形地貌，保留自然生长的植被，招引小鸟，成为鸟类生活的天堂。由于湿地小岛在洪水期会被淹没，因此不设置固定游览设施，仅适当开辟游览小路，游船停靠点，供游人观光游览。

在组团内部，建筑周围设置细腻、精致的庭院园林，运用花架、假山、雕塑、小品、花坛、流水，强调组团入口特色。园林绿化特色与建筑风格特色呼应，使每栋建筑有最佳的视线效果，让人们生活在安宁与悠闲的环境当中。

KEYWORDS 关键词

Exquisite Waterscape
精致水景

Palm Plants
棕榈植物

Different Sceneries by Moving
移步异景

Southeast Asian Style
东南亚风格

Location: Zhongkai Hi-Tech Zone, Huizhou, Guangdong
Landscape Design: Guangzhou Homy Landscape Co, Ltd.
Landscape Area: 56,748 m^2

项目地点：广东省惠州市仲恺高新区
景观设计：华誉国际景观设计公司
景观面积：56 748 m^2

Guang Group · Valley in City
光耀 · 城市山谷

FEATURES 项目亮点

The use of distinctive landscape wall, cascade, pools and other elements, togethered with the naturally scattered tropical plants, it integrates the natural landscape and man-made landscape.

充分利用具有鲜明特色的景墙、叠水、泳池等元素，配合自然错落的热带植物，让自然景观与人造景观融为一体。

Overview

The project located in the back of Huihuan Street Office, Zhongkai High-tech Zone in Huizhou City, next to Zhongkai Avenue, adjacent to exit of Dongguan-Huizhou light rail, is a real superstructure above light rail. With planning land area of about 500,000 m^2, surrounded by mountains on three sides, it is another large project of mountain resource developed by Guang Group, all kinds of facilities will be improved successively in the future. The phase I and II cover land area of 80,000 m^2 and floor area of about 120,000 m^2.

项目概况

光耀·城市山谷位于惠州市仲恺高新区惠环街道办背面，毗邻仲恺大道，紧邻莞惠轻轨出口，是真正的轻轨上盖物业。项目拟规划用地约 50 万 m^2，三面群山环抱，这将是光耀集团开发的又一山景资源大盘，届时各种配套设施将会陆续完善。其中一、二期占地 8 万 m^2，建筑面积约 12 万 m^2。

Design Objective

By understanding and innovation of the Southeast Asian style, combined with unique geographical location and height change, the project forms urban forest landscape. Exquisite detail landscape sketch fully enhances the overall landscape's cultural depth. The overall landscape spatial layout expands garden culture and space tension, and creates romantic, warm and harmonious community atmosphere, reproducing a kind of leisure, gentle and natural humanistic environment of the Southeast Asian style.

设计目标

通过对东南亚风格的理解与创新突破，结合项目独特的地理位置和高差变化，项目形成城市山林园林景观。通过精致细部景观小品提升整体园林文化深度，通过整体景观空间布局营造扩展园林文化与空间张力，营造了一个浪漫、热情、和谐的社区氛围，再现一种休闲写意、亲切自然的东南亚风情的人文环境。

Landscape Planning

The landscape design goal, taking the most wonderful waterscape in Southeast Asia as the soul in the noisy city space, is to make full use of distinctive landscape wall, cascade, pools and other elements, together with the naturally scattered tropical plants, to integrate the natural landscape and man-made landscape and to build a delicate waterscape garden town of Southeast Asian style, in which people enjoy the leisure life away from the hubbub.

The plot's plant design is in Southeast Asia style with rich plant species and distinct gradation, and mainly uses the regional featured palm plants, matched with the flowering trees and evergreen trees. Plant form combines geometric and natural style, to create Southeast Asian garden landscape among modern buildings. Each movement of step will bring the different sceneries, embodying leisure life of the community.

景观规划

景观设计根据目标，在烦嚣的城市空间之中，以东南亚景观中最精彩的水景作为灵魂，充分利用具有鲜明特色的景墙、叠水、泳池等元素，配合自然错落的热带植物，让自然景观与人造景观融为一体，打造一座东南亚风情的精致水景园林小镇，令生活在其中的人们享受远离尘嚣的休闲生活。

本地块植物设计属东南亚风格。植物种类丰富，层次分明，以具有东南亚特色的棕榈类植物为主，配以开花、常绿乔木。栽种形式采用几何式与自然式相结合，在现代化的建筑物之间营造东南亚园林景观，让人移步异景，体现休闲的社区生活。

KEYWORDS 关键词

Concise
简洁明快

Leisure
休闲情趣

Meandering and Secluded
曲径通幽

Southeast Asian Style
东南亚风格

Location: Chikan District, Zhanjiang, Guangdong
Developer: Zhanjiang Kingkey Real Estate
Landscape Design: Botao Landscape (Australia)
Landscape Area: 70,000 m^2

项目地点：广东省湛江市赤坎区
开发商：湛江市京基房地产开发有限公司
景观设计：澳大利亚・柏涛景观
景观面积：7 万 m^2

Western Guangdong Kingkey City
西粤京基城

FEATURES 项目亮点

Landscape at both sides of the central axis is equally arranged. The left side adopts the landscape theme of static water surface and floating wood platform, to depict the overall leisure sense of the development; while the right side is designed with modular planting and sloping lawns to form variety of landscape layers.

中心轴左右均衡布置，左侧以静水面和漂浮在水上的木平台为景观主题，刻画小区整体的休闲情趣；右侧为整齐树阵和坡地草坪，层层叠叠摇曳多姿。

Overview

Western Guangdong Kingkey City is located in People's Avenue South, across from Zhanjiang City Sports Center, which is the two cores—Central Business District (CBD) and Central Living District (CLD). The project covers land area of 280,000 m^2, total floor area of more than 1,000,000 m^2, erecting in the central axis of Zhanjiang City, famous as CBD landmark.

项目概况

西粤京基城位于湛江市人民大道南，湛江市体育中心对面。项目地处湛江城市双核——中央商务区（CBD）、中央居住区（CLD）。项目占地面积 28 万 m^2，总建筑面积超过 100 万 m^2，傲然屹立于湛江城市中轴线上，席揽 CBD 地标之尊。

Aerial View 鸟瞰图

Design Objective

Landscape design continues the architectural planning features, and the overall style is the modern Southeast Asian seaside landscape. In the overall layout of the landscape design, based on the three courtyards formed by architectural planning, different treatment of landscape site creates different landscape experience spaces, meanwhile the continuation and connection of landscape axis makes the three courtyards interrelate and mutually penetrate. The overall design of the community, based on man-made natural landscape design, emphasizes on creating landscape axis and landscape center. Road axis and water systems connect to form a whole space.

设计目标

景观设计延续建筑规划特色，整体风格定位为现代东南亚海滨景观。在景观整体布局设计上，利用建筑规划所形成的三个庭院，有区别地对待景观场地，营造不同的景观体验空间，同时也利用景观轴线的连接与延续，使三个庭院相互关联、互为渗透。小区整体以人工化的自然景观设计为基础，强调景观轴线与景观中心的营造，利用道路轴线和水系串联，形成一个整体空间。

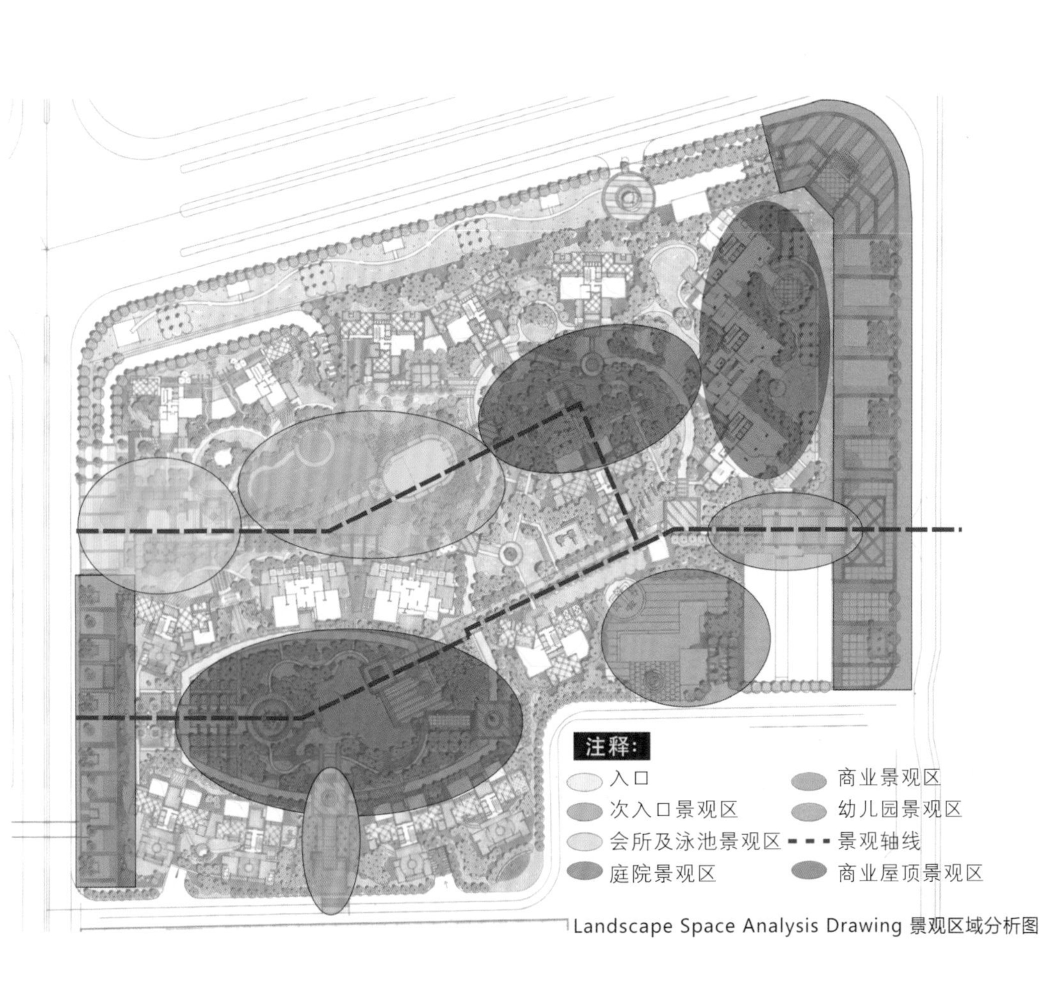

Landscape Space Analysis Drawing 景观区域分析图

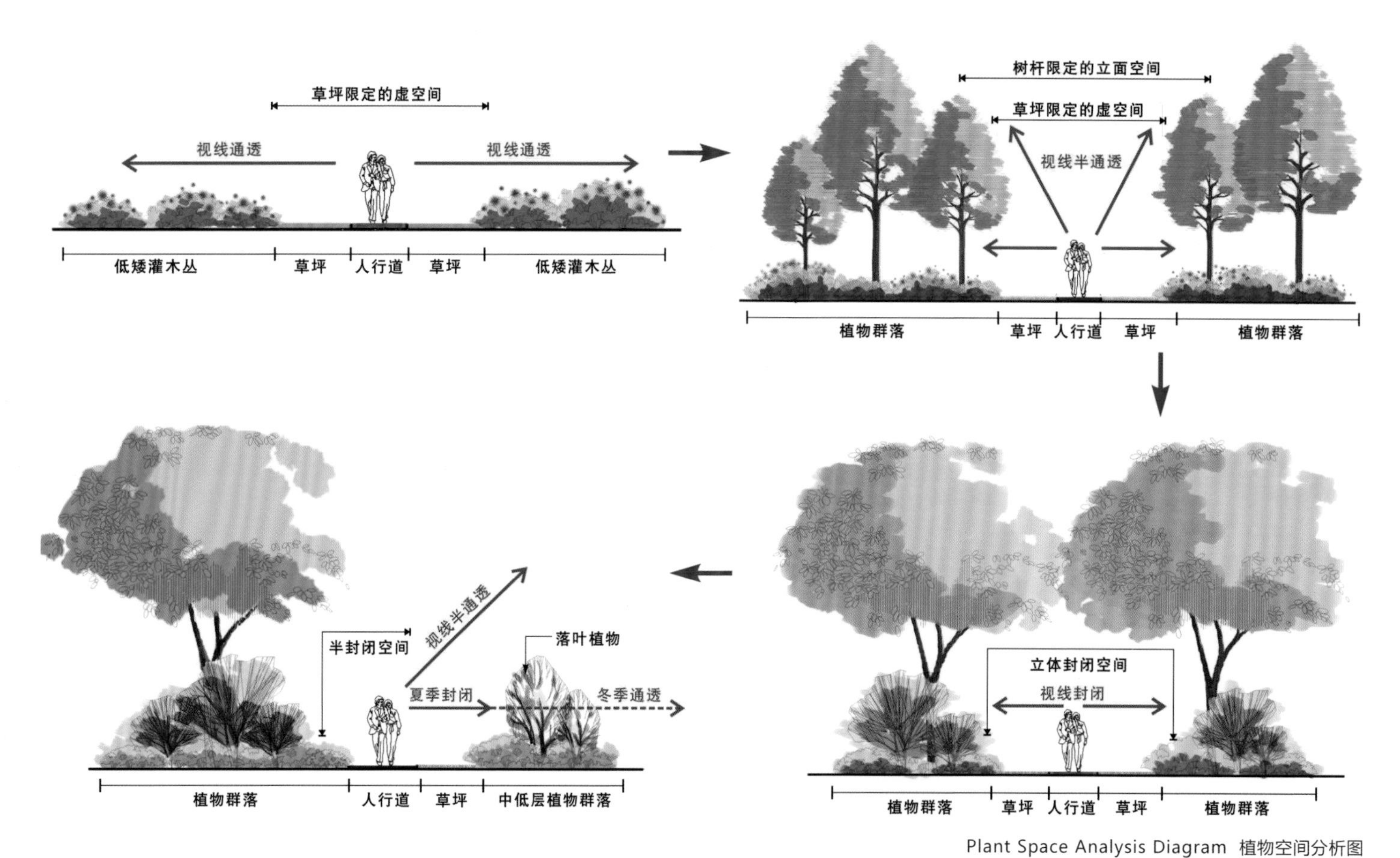

Plant Space Analysis Diagram 植物空间分析图

Landscape Planning

Main entrance is the window of showing the building's landscape taste, so it adopts the half-open landscape layout in design to show out more landscape planes. Landscape at both sides of the central axis is equally arranged. The left side adopts the landscape theme of static water surface and floating wood platform, to depict the overall leisure sense of the development; while the right side is designed with modular planting and sloping lawns to form variety of landscape layers.

The second courtyard takes streams and green lands as the main elements. Natural mottled light and shadow bring wonderful feeling; green space and water system complement each other and form a delightful contrast. The winding trail under the trees is meandering and secluded, and its geometric layout is elaborate though appears random. The strip-like arc road goes across the whole courtyard. Geometric and crisscross paths are arranged to divide landscape function space. Rest area where delicate pavilion and gallery or vivid pieces are dotted emphasizes the sense of home by planting colorful flowers and leaves of shrubs lush around.

The third courtyard uses regular geometric layout, composed by two big enclosed arcs to make once irregular buildings neat and orderly. Rows of trees and free pools create a simple and concise courtyard space. A large area of rest square provides multiple venues for old people doing morning exercises, adult socially communicating and children playing. At night, the colorful lights create light and shadow, virtual and real spiritual space.

景观规划

主入口是整个楼盘对外展示景观品味的窗口，在景观设计上，采用了半开敞式景观布局，以便把更多的景观面对外展示。中心轴左右均衡布置，左侧以静水面和漂浮在水上的木平台为景观主题，刻画小区整体的休闲情趣；右侧为整齐树阵和坡地草坪，层层叠叠摇曳多姿。

第二庭院以小溪和绿地为主要构成元素。自然斑驳的光影，带给人奇妙的感受；绿地与水体相辅相成，相映成趣。树下蜿蜒的园中小径，曲折幽静，较为自由的几何布局看似随意，实则精心。以弧线的带状园路串起整个庭院，利用纵横交错的几何小径划分景观功能空间，休息处或设精致亭廊架，或设动人小品，周边遍植花叶繁茂的各类灌木，色彩丰富，强调居家之感。

第三庭院采用规则几何布局，两个大线条弧度的围合构成，使得原本不太整齐的建筑围合空间整齐有序。成排的树阵、自由的水池，营造出一个简洁明快的庭院空间。大面积的休息广场作为提供给老人晨练、大人社交、小孩嬉戏的多功能场地。夜间，缤纷的灯火营造出光与影、虚与实的灵性空间。

KEYWORDS 关键词

Indonesian Style
印尼风情

Colorful
色彩缤纷

Tropical Flavor
热带气息

Southeast Asian Style
东南亚风格

Location: Nanning, Zhanjiang, the Guangxi Zhuang Autonomous Region
Developer: Guangxi Greatwall Real Estate Development Co., Ltd.
Landscape Design: Botao Landscape (Australia)
Landscape Area: 30,000 m^2

项目地点：广西壮族自治区南宁市
开发商：广西长城房地产开发有限公司
景观设计：澳大利亚・柏涛景观
景观面积：3 万 m^2

Nanning, Indonesian Garden
南宁印尼园

FEATURES 项目亮点

Garden design based on "people-orientation", with reasonable and rigorous landscape layout, makes people feel being in authentic Indonesian style garden.

园林设计上“以人为本”，景观布局合理而严谨，让人彷如置身于原汁原味的印度尼西亚风情园林中。

Overview

Nanning, Indonesian Garden is located in the intersection of Guiya Road and Hezuo Road, extending north to the main traffic road of Minzu Road, west to Shimen Forest Park, south to Qingxiushan Park and sports theme park, and east to Park for ASEAN countries. The project takes full advantage of the slope's height difference, effectively separates the commercial building, office building and residential building, and blends the quietness of residential area with the passionate feel of commercial office area in harmonious way by virtue of luxury and characteristic landscape garden in the Bali hotel.

项目概况

南宁印尼园位于桂雅路与合作路交汇处，北临民族大道交通主干道，西临石门森林公园，南靠青秀山公园和体育主题公园，东侧为东盟各国园区。项目充分利用坡地高差，将商业、写字楼与住宅进行有效的区隔，并借助印尼巴厘岛酒店式奢华特色山水园林，将居住区的宁静与商业办公区的风情洋溢以和谐的方式调和在一起。

印度尼西亚之家
HOUSE OF INDONESIA

Site Plan 总平面图

Design Objective

Garden design based on "people-orientation", with reasonable and rigorous landscape layout, makes people feel being in authentic Indonesian style garden.

设计目标

园林设计上“以人为本”，景观布局合理而严谨，让人彷如置身于原汁原味的印度尼西亚风情园林中。

Main Entrance Elevation 主入口立面图

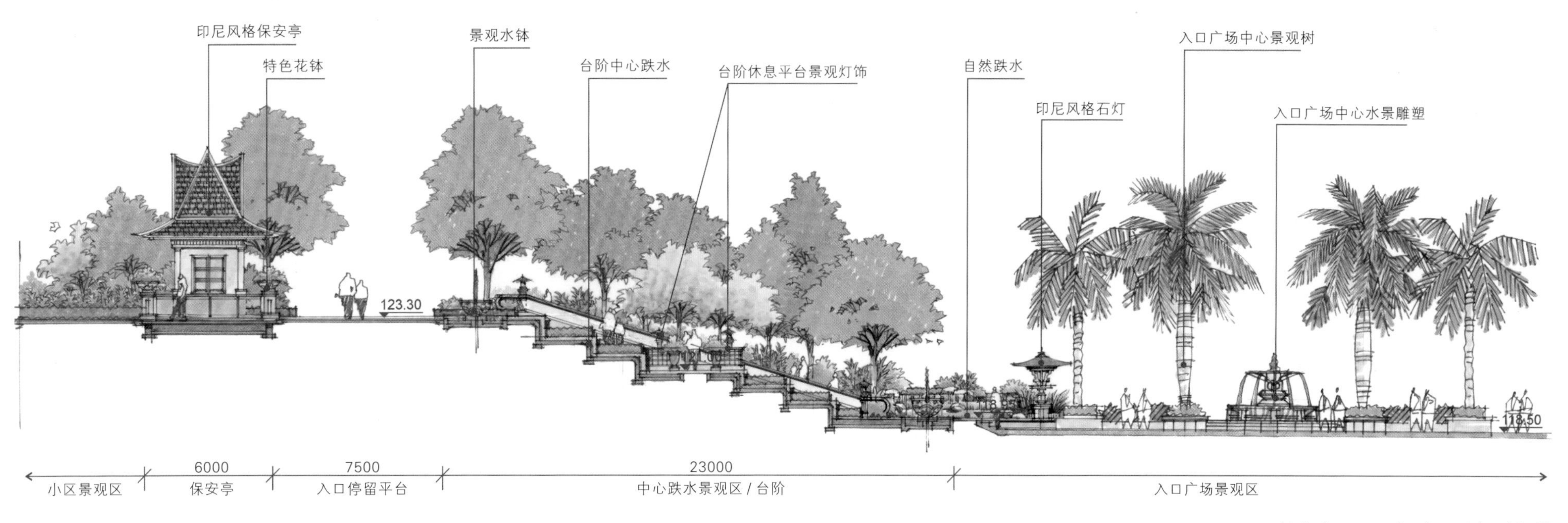

Main Entrance Section 主入口剖面图

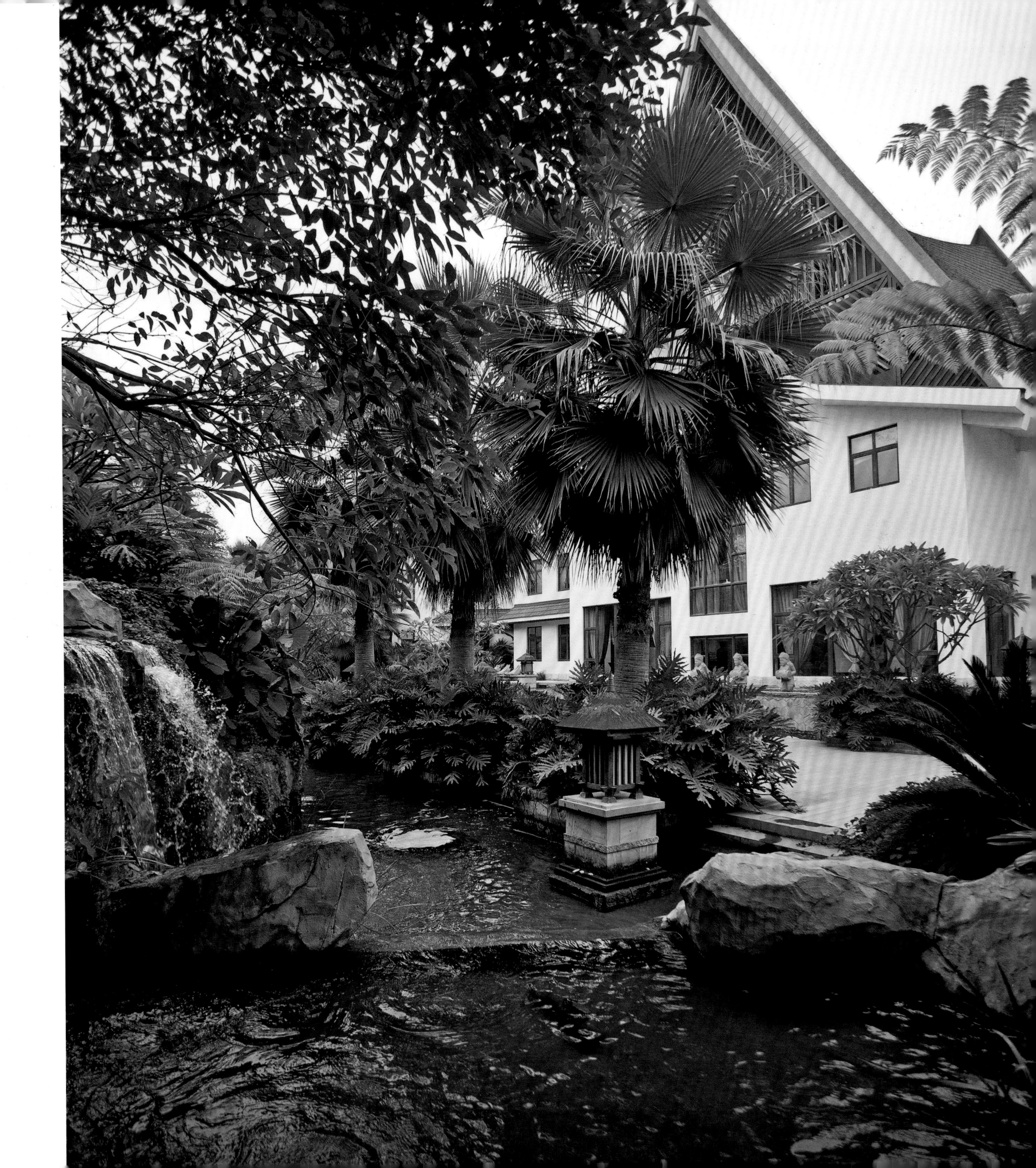

Landscape Planning

Nanning Indonesian Garden transfers the most real beauty and charm from Indonesia. Garden design is based on people-orientation. The best garden design can have different spaces and region providing useful functions for users at all ages. Lush planting dotted with colorful petals is arranged naturally in the garden. At the same time, the garden structures such as landscape wall full of rough texture, pavilions in exquisite modelling, wood gallery frames full of tropical customs, landscape bridges, sculptures, wooden platforms, the characteristic porch and short lamp holders in the entrance door, all express strong tropical flavor from Indonesia. The typical landscape conception of Indonesia traditional gardens in the famous Bali Island is reflected in the project.

Central landscape, located in the heart of the Garden, with the largest flow of people, has the appeal of maximizing landscape features. The area includes dynamic and bustling modern commercial street. Plant design in the garden is cut into stair-step shape, layer upon layer. This design inspiration is derived from the Indonesian native countryside terraces. Similar design details inspired from Indonesian unique scenery can be seen everywhere. Featured waterscape is located in the heart of the Garden. The flowing water system brings vitality, extending from the center to the surrounding yard, till to the two featured water attractions placed in commercial street.

景观规划

南宁印尼园传递的是来自印度尼西亚最真实的美景与韵味。园林设计以人为本，通过最好的园林设计，使其拥有不同的空间和区域，为不同年纪的使用者提供所用的功能。茂密的种植以最自然的方式布置在园林中，配以色彩缤纷的花瓣点缀其中。同时，充满粗犷肌理的特色景墙，造型讲究的凉亭，充满热带民俗风情的木廊架、景桥、雕塑和木平台，还有入口大门特色门廊、园林矮灯座等园林构筑物，都传达着来自印度尼西亚浓厚的热带风情。这些在著名的巴厘岛典型印尼传统园林中的景观意境都在项目中有所体现。

中心景观位于园区的核心位置，也是人流量最大的地方，具有最大化特色景观的感染力。该区域有洋溢着热闹动感氛围的现代商业街。园林中的植物设计被修剪成阶梯状，层层相叠。这样的设计灵感正是源于印尼本国乡间的梯田，类似这样源于印尼独有风景的灵感设计细节随处可见。特色水景位于园区的中心地带，流动的水系给园区带来了生命的活力，从中心向院里的四周延伸，直流到商业街设置的两处特色的水景点。

KEYWORDS 关键词

Ecological Tree Belt
生态林带

Urban Oasis
都市绿洲

Tropical Plants
热带植栽

Southeast Asian Style
东南亚风格

Location: Foshan, Guangdong
Client: Foshan SUNUNI
Landscape Design: Keymaster Consultant (Guangzhou) Ltd.
Design Team: Cai Shuyan, Zhu Lei, Peng Yi,
Feng Zhuowei, Zhao Pengfei
Landscape Area: Approx 50,000 m^2

项目地点：广东省佛山市
业主：佛山兆阳地产
景观设计：广州市科美都市景观规划有限公司
设计团队：蔡舒雁、朱蕾、彭毅、冯焯伟、赵鹏飞
景观面积：约 5 万 m^2

Landscape Design for SUNUNI Royal Garden, Foshan

佛山兆阳御花园景观设计

FEATURES 项目亮点

The main landscape creates two green belts of trees which run through the whole community to bring the residents with sustainable landscape enjoyment. Consummated road systems are designed along the two green tree belts, enabling people to wander in the "forest", breathe with green and enjoy the relaxing outdoor lifestyle in this "urban oasis".

主体景观上创造两条贯穿整个社区的生态绿色林带，为项目带来可持续的景观享受。根据两条生态林带设置完善的道路系统，犹如漫步森林，每天与绿色一同呼吸，让人真正享受到“都市绿洲”这一轻松的户外生活方式。

Overview

Located on Huayuan East Road, Chancheng District of Foshan City, the complex is composed by high-rise residential buildings and commercial facilities, enjoying convenient transportation and complete service facilities.

项目概况

项目位于佛山市禅城区华远东路，建筑群由高层区和商业区两部分组成，交通便捷，服务设施配备完善。

北陽御花園

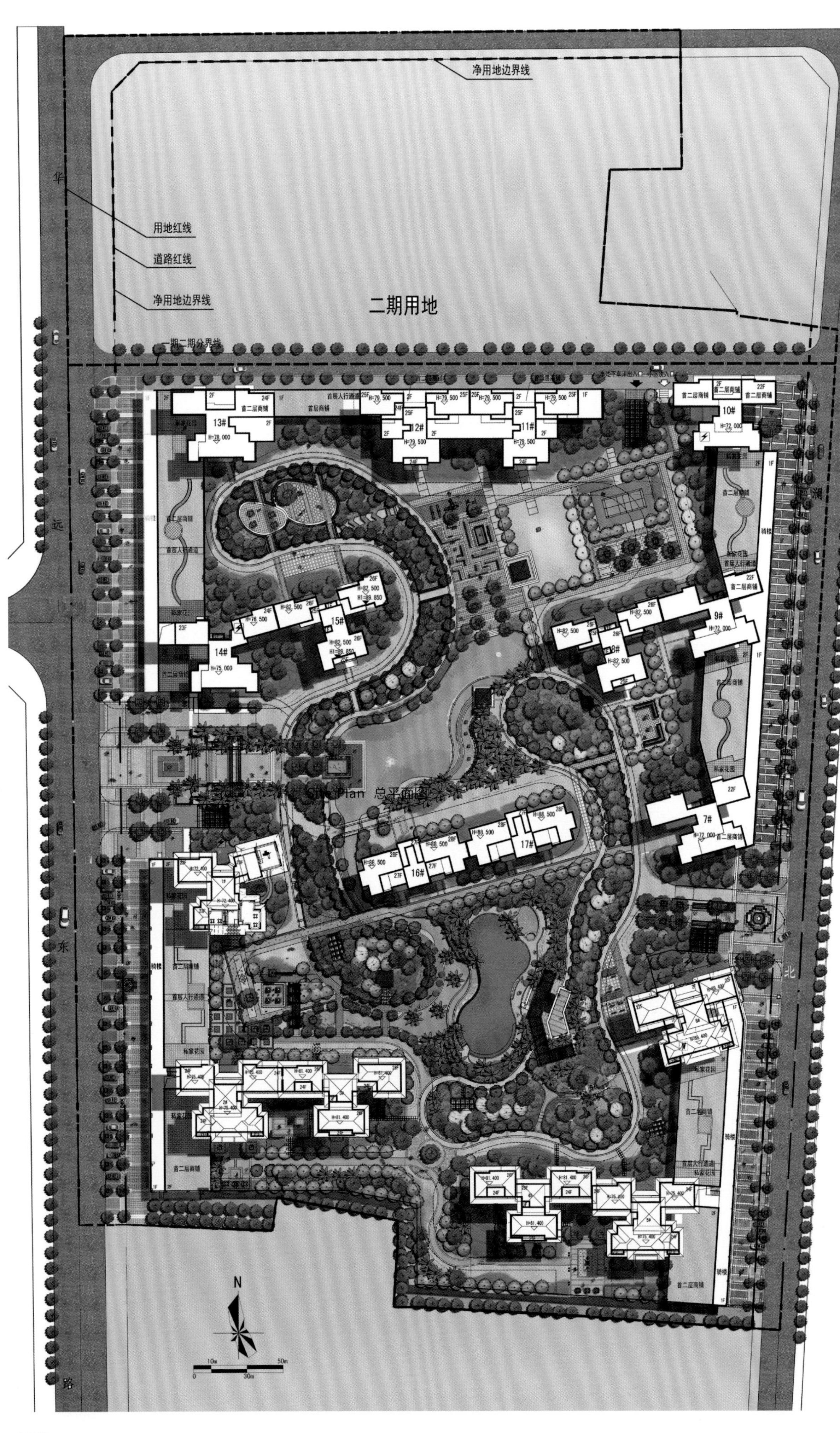

Landscape Design

The project boasts enclosed-layout with strong sense of territory. The overall planning inclines up to 15°, which has avoided the stiffness of conventional layout and shaped the identity for the landscape. The main landscape creates two green belts of trees which run through the whole community to bring the residents with sustainable landscape enjoyment. Consummated road systems are designed along the two green tree belts, enabling people to wander in the "forest", breathe with green and enjoy the relaxing outdoor lifestyle in this "urban oasis". Under the limited conditions, the designers have tried their best to create diversified topography and defined the green tone of the project with two ecological mountains. Furthermore, an eco river system is designed to match the mountains and create an environment with both water and mountain.

The lifestyle of "urban oasis" emphasizes the creation of outdoor living atmosphere. This idea has been fully considered during the landscape design. By the combination of two eco belts of trees, the residents and visitors will truly experience this kind of relaxed outdoor lifestyle.

From the basic parts to the detail elements, the designers have enabled people to have a great experience of the oasis lifestyle in modern city by every detail. The ecological tropical plants are arranged in traditional and classical style to match the architectural style and emphasize the unique style of the project. The main entrance is designed magnificently and symmetrically with dynamic water features, elegant colors, exaggerated iron artworks and appropriate plant to be an important symbol of the project.

The project has interpreted traditional and classical landscape art with modern techniques. The hard landscape made of beige stones is decorated by some mysterious and elegant black glazed stones. In addition with the exquisite iron decorations, the warm-toned wooden platforms and the trestle bridges by river, every landscape node is well planned and designed with unique characteristics. Considering the target users of the development, the designers pay attention to the lighting effect at night: the basic lighting system is skillfully designed to match different scenes and enhance the artistic atmosphere. For the key landscape buildings and items, the designers carefully employ varied lighting skills to enhance the outline, to point out the center, to make the waterscape more dynamic, or to highlight the tranquility of the environment. The lighting effect will match the surroundings perfectly to create a beautiful and comfortable space for nighttime activities.

景观设计

项目总体采用围合式布局，具有强烈的领域感。规划布局上倾斜 15°，打破了常规布局的呆板，为景观的个性营造创造了有利条件。主体景观上创造两条贯穿整个社区的生态绿色林带，为项目带来可持续的景观享受。根据两条生态林带设置完善的道路系统，犹如漫步森林，每天与绿色一同呼吸，让人真正享受到“都市绿洲”这一轻松的户外生活方式。在有限的条件下，最大化地创造出丰富的地形变化，两条生态山脉成为整个项目的绿色基调。此外，生态水系和“山脉”共同打造“山水相依”的规划格局。

“都市绿洲”的生活方式着重于户外的居住氛围的营造，这一理念强烈地贯穿于整个景观设计之中。通过两条生态林带的组合贯通，使居民和来访者真正享受到这一轻松的户外生活方式。

从最基本的局部到细部元素的设计，现代都市居住和轻松休闲的绿洲生活方式在每一个细节中都体现得淋漓尽致。生态的热带植栽与传统型古典及设计的融合，与建筑相呼应，更展示了这一项目的独特风格。小区主入口的设计很好地反映了这一理念，规整对称的布局、灵动的水景、淡雅的色调、浮夸的铁艺，再加上植物的配搭，整个场景大气而内敛，成为项目的一个重要的标志。

项目整体以现代手法诠释传统古典的景观艺术，硬景的用材以米黄色石材为主，用神秘高雅的光面黑色石材为点缀，加上精致的铁艺装饰，另外用暖色调的木平台、栈桥等与水系结合，使每个节点都统一而各具特色。考虑到项目服务的对象，灯光夜景效果也是设计的重点，基本的照明功能结合场景的布局，使灯光融入景观艺术之中。对于主要的景观单体和小品，设计师通过反复推敲，运用了多种灯光表达手法，或是强化小品的轮廓，或是突出场景的中心，或是使水景更加富有动感，或是衬托环境的幽静。灯光与环境的完美结合，为住户提供了一个优美舒适的夜间活动场所。

Vegetation Design

The palm plants such as the Phoenix Sylvestris, Washingtonia Filifera, Wodyetia Bifurcata and Arenga Tremula are employed to enrich the vegetation skyline and enhance the tropical style. At the same time, the design has emphasized the ecology of the vegetation by using local species like Cassia surattensis, Bauhinia blakeana, Bischofia javanica and Alstonia scholaris, achieving a balance between ecology and landscape views.

植被设计

银海枣、老人葵、狐尾椰子及鱼骨葵等棕榈科植物不仅极大地丰富了植栽的天际线，也渲染了浓烈的热带风情；同时项目也强调栽植的生态性，大量采用黄槐、红花紫荆、秋枫及盆架子等乡土树种，满足了生态性和景观性的平衡。

KEYWORDS 关键词

Waterscape Wall

水景墙

Palm Tree Array

棕榈树阵

Cascades and Waterfalls

跌水景观

Southeast Asian Style

东南亚风格

Location: Putian, Fujian

Developer: Zhenro Group

Landscape Design: PASNO Landscape Architecture Studio

Land Area: 30,843.92 m^2

项目地点：福建省莆田市

投资方：福建正荣集团

景观设计：普梵思洛（亚洲）景观规划设计事务所

占地面积：30 843.92 m^2

Zhenro · Luxury Mansion Blue Bay, Putian

正荣莆田御品兰湾高档住宅区

FEATURES 项目亮点

The design approach of transition between opening and closing is to create a courtyard landscape of different views as one moves and to infuse the ambiance of Thai-styled landscapes into the buildings. Based on the Southeast Asian style, the whole project has well interpreted the pure flavor of the Phuket Island, the Pattaya and the Samui Island of Thailand, showing the rich exotic flavor.

本项目以开合转折的设计手法，营造出移步异景的庭院景观，把泰式景观的韵味渗透在楼宇之间。整个项目以东南亚风情园林为蓝本，将纯粹的普吉岛风情、芭堤岛风情以及泰国的苏美岛风情原味演绎，体现饱满的异域风情。

Overview

The project is located in an important development area in Putian's "one stream and two banks" region— Mulanxi Area which will integrate business, commerce, residences, tourism, culture and entertainment, providing people with the civil entertainment center, commercial complex, headquarters base, waterfront theme park, five-star hotel, boutique shopping street, art and culture center, and cultural innovation park. It will greatly increase Putian's value and drive its economic development.

项目概况

项目位于莆田“一溪两岸”重点开发区域——木兰溪板块，伴随着海西发展和莆田城市发展战略的推进，木兰溪区域将集商务、商业、居住、度假、文化、娱乐等于一体，涵盖市民文化娱乐中心、商业综合体、总部办公基地、滨水主题公园、五星级酒店、精品购物水街、艺术文化中心和文化创意园等“世界级”配套，将引领莆田的价值高地和经济巅峰。

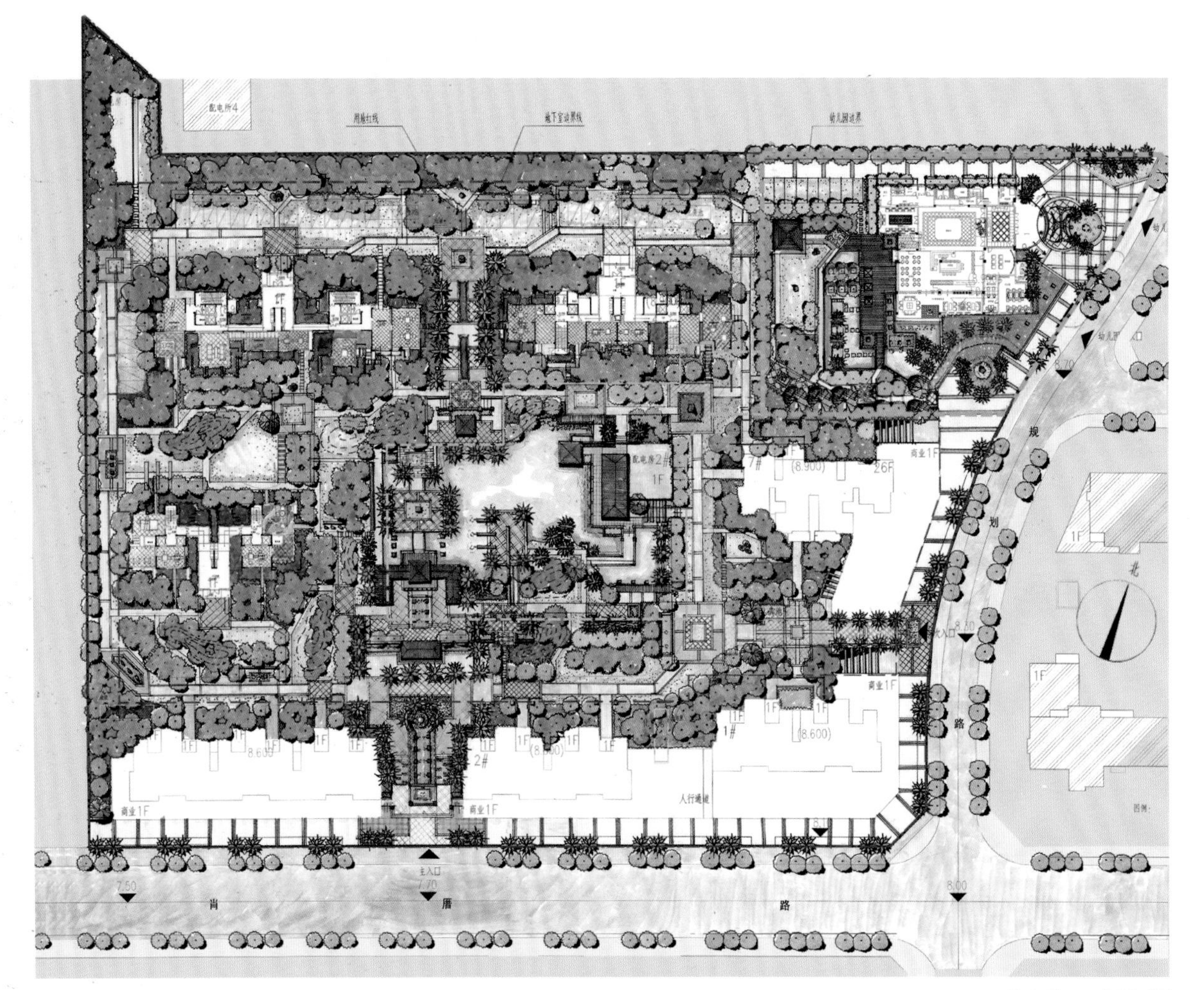

Site Plan 总平面图

Design Idea

How to explore the characteristics and advantages of a land, meet the requirements of the market, and define the style of the landscape, is the major challenges for the landscape designers. By carefully analyzing on the site conditions, the designers of this project have thought that the landscape needs to be exquisite, ecological, pure and exotic to match the architectural style and shape the living culture of Putian City. Therefore, the designers have proposed the idea of "inhabiting in nobility, hiding in a foreign land", and defined the landscape style as the exotic Thai style. Based on the Southeast Asian style, the whole project has well interpreted the pure flavor of the Phuket Island, the Pattaya and the Samui Island of Thailand, showing the rich exotic flavor .

设计理念

如何发掘项目地块的特征及优势，瞄准市场，准确导入景观风格是景观设计的重点所在。设计师经过对项目地块认真分析，认为项目要走精致、生态、纯粹的异域风情路线，用异域风情装点建筑环境，打差异化牌，引领莆田居住文化。于是设计师提出“栖居于尊贵，私隐于异乡”的理念，景观设计风格定位为异域风情：泰式园林。整个项目以东南亚风情园林为蓝本，将纯粹的普吉岛风情、芭堤岛风情以及泰国的苏美岛风情原味演绎，体现饱满的异域风情。

Overall Landscape Design

The buildings are high rises arranged in a circle to form an enclosed landscape center. With skillful and rhythmic design, it has created a comfortable space for leisure and relaxation. The rooftop of the underground garage covered by 1-1.5 m thick earth is fully used to provide more landscape space. At the same time, to increase space levels on this flat site, varied terraces, lush plants, changeful waterscapes are employed to create a colorful landscape space.

The design approach of transition between opening and closing is to create a courtyard landscape of different views as one moves and to infuse the ambiance of Thai-style landscapes into the buildings. The stylish landscape pavilions and corridors connecting the exquisite crystal spaces, together with the carefully selected Thai-style sculptures, flower pots, small decorations, have created a garden space which is superior in quality. At the same time, the carefully designed vegetation has soften the architectural spaces and provided people with green activity spaces. This kind of soft landscape has helped to create a series of comfortable, pleasant, ecological and functional spaces.

总体景观设计

整个项目建筑为高层，且成环形分布，围合感较强，充分利用组团中心景观带，通过精巧、有节奏的设计，打造舒适的休闲空间，充分利用地库顶板 1~1.5 m 的覆土深度，设计出丰富的空间变化，同时土地现状较为平整，缺少空间层次的变化，通过错落的台地、饱满的植物、变幻的水景来打造丰富多变的景观空间。

本项目以开合转折的设计手法，营造出移步异景的庭院景观，把泰式景观的韵味渗透在楼宇之间。风情感十足的景观构筑亭廊，穿梭于精致的水景空间且经过精心挑选的泰式雕塑、花钵、小品，营造出品质感十足的园林空间，小区的高端景象尽收眼底，将最精彩的景观空间展示在人们面前。同时设计师以悉心营造的植栽空间软化建筑的生硬，在一片绿色间将活动功能置于其中，以软景打造宜人、舒适、生态的实用空间。

质量是企业的生命
第一团队

10F
2F

8F
6F
4F

Landscape Areas Design

Central Courtyard — Landscape Elements: characteristic landscape architectures, sculpture fountains, flower pots, wooden viewing platform, waterside steps, wooden plank road, landscape wall with fountains, springs, waterfalls, palm tree array, landscape corridor and micro-topography. The central waterscape creates a comfortable waterfront environment, and the elegant leisure platform and natural water features contribute to a noble lifestyle.

Areas Between Building Groups — Landscape Elements: landscape corridor, leisure square, characteristic landscape items, landscape trees, micro-topography, landscape wall and leisure seats.

Temporary Sales Center — Landscape Elements: waterscape wall, characteristic landscape items, springs, pavements, wooden platform, micro-topography and the stone steps over water. Here all the landscape elements are exquisitely designed to highlight the style of the development and create a viewing space of five-start hotel style. It shows the respect to nature, health, leisure style and local customs, trying to integrate the landscape into the surroundings.

Commercial Area and Entrance Area — Landscape Elements: leisure seats, parasol, lamp poles, planting pools and tall palm trees. The landscape axis at the main entrance fully showcases the high quality and the integral style, highlights the nobility of life, and enhances the exotic style of the project.

各景观区域设计

中心庭院景观——造景元素：特色景观构筑物、喷水雕塑、花钵、观景木平台、亲水台阶、木栈道、喷水景墙、涌泉、跌水景观、棕榈树阵、景观廊架和微地形。中心水景动静结合营造舒适的水岸生活，雅致的休闲平台、自然而亲切的水景，酝酿尊贵的生活。

组团景观区域——造景元素：景观廊架、休闲广场、特色景观小品、特色景观树、微地形、景观坐墙和休闲座椅。

临时售楼区域——造景元素：景观水景墙、特色景观小品、涌泉、特色景观铺装、木平台、微地形和汀步。走精致的景观路线，展示楼盘的风格基调，营造五星级酒店式的观景空间，崇尚自然、健康、休闲的特质及风土人情，与周边的环境融为一体。

商业及入口区域——造景元素：休闲座椅、阳伞、特色景观灯柱、景观种植池和高大的棕榈树。主入口景观轴线作为小区对外界面，充分展示高端品质与整体风格，突出生活的尊贵感，让楼盘充满着异域风情。

逾万平米泰式皇家

KEYWORDS 关键词

Mountain Villas Style
山居风情

Stereo Ecology
立体生态

Distinct Layers
层次丰富

Southeast Asian Style
东南亚风格

Location: Zengcheng District, Guangzhou, Guangdong
Landscape Design: Guangzhou Homy Landscape Co., Ltd.
Land Area: 123,333 m²

项目地点：广东省广州市增城区
景观设计：华誉国际景观设计公司
占地面积：123 333 m²

Jiewei · Eastern Mansion
杰伟 · 尚东紫御

FEATURES 项目亮点

The gardens in Southeast Asian style with distinct layers and unique water features have followed the characteristics of nature, health and leisure, making full use of interspersion of plants to represent fully respect for nature and pursuing for harmony between human and nature.

东南亚风情园林，承续自然、健康和休闲的特质，层次分明、特色水景，充分利用景观的绿意点缀，体现了对自然的尊重和对人居的崇尚。

Overview

Nestled in Guangzhou sub-center of Licheng, Zengcheng, this project is the core of three towns that are Gualv New Town, Education Town of Zhu Village and Zhongxin Knowledge Town, with brand new western style house groups, while delicate landscape view are fully employed in the development, and unique private yards are introduced to the houses to create pleasant lives that are far away from dirt but close to the urban center. Development has the unit types of 89 m² for three bedrooms and 109 m² for four, both with high efficiency ratio and private yards.

项目概况

尚东紫御位于广州城市副中心——增城荔城，地处挂绿新城、朱村教育城、中新知识城三城核心，全新禧园洋房组团，精心打造立体园林景致，步步皆景，个性写意的入户花园，将风景筑在家里，真正达到离尘不离城的生活。89 m² 三居室、109 m² 四居室，超高实用率，户户带入户花园。

Design Objective

In order to express the dignity and nobility of the development, landscape design starts from improving the elegant taste by using the methods that for luxury villas to represent sumptuousness and grace of living spaces. In the meanwhile, designers make fully use of slope and height differences, as well as landscape elements of hillside, mesa, valley, lake and so on, to create pure and elegant landscape characteristics and to enhance it as ecological mountain landscape, which have met even exceeded the requirements for landscape of targeted residents.

设计目标

为体现紫御大宅的尊贵与高度，园林景观设计拟从提升大宅尊贵品味方面着手，用别墅手法营宅，以表征人居空间的奢华与优雅。项目地势为坡地，景观设计致力于巧妙地运用地形、高差等特点，通过山坡、台地、溪谷、湖泊等景观元素，形成纯粹、高贵的景观特色，强调“山居立体式生态”景观，满足又高于目标消费群体对园林景观的要求。

Landscape Planning

Making the space usage to the maximum skillfully, designers pay a lot attention on the comfort and daylighting design of spaces, as well as colorful spatial levels, in order to be full of creativity but also environment friendly. The gardens in Southeast Asian style with distinct layers and unique water features have followed the characteristics of nature, health and leisure, making full use of interspersion of plants to represent fully respect for nature and pursuing for harmony between human and nature. Gardens that set at front and back of villas are cooperating with landscape on two sides as well as terraces on the ground and second floor to create stereo spaces, residents are able to enjoy the fun of gardening and the beauty of surrounding landscape.

景观规划

设计师巧妙地将空间使用最大化，在设计上，既有创新更不失环保，空间层次丰富，在空间舒适度和采光设计上花足心思。东南亚风情园林，承续自然、健康和休闲的特质，层次分明、特色水景，充分利用景观的绿意点缀，体现了对自然的尊重和对人居的崇尚；前后花园、左右园林及上下露台配比的立体空间，可得养草弄花之乐，也可享花园林景之美。

KEYWORDS 关键词

Natural and Ecological

自然生态

Exotic Style

异域风情

Artistic Decorations

艺术小品

Southeast Asian Style

东南亚风格

Location: Quanzhou, Fujian
Developer: JOINV Group
Landscape Design: PASNO Landscape Architecture Studio

项目地点：福建省泉州市
开发商：卓辉地产集团
景观设计：普梵思洛（亚洲）景观规划设计事务所

JOINV the Gold Bund

卓辉泉州金色外滩滨江高档社区

FEATURES 项目亮点

The design highlights the integration of the natural environment and the Bali style, and the complementary design concept between the natural elements and luxurious architectures. It adopts the tropical style of the paradise resort and simplicity and comfort of Bali Islands to create a Garden of Eden in the city which features exotic style and cozy holiday atmosphere.

设计体现天然环境与社区巴厘岛风情的互融、自然元素和豪华构筑互补的设计理念，运用巴厘岛度假天堂的热带风情及巴厘岛的质朴闲适，营造出含异域风情、惬意高雅的度假气氛，打造城市中的梦幻伊甸园。

Overview

Occupying a total area of 80,000 m^2, the project is located in the northwest of Quanzhou, on the west bank of Jinjiang River to enjoy great river views. There are high-rise buildings, medium-rises and multi-storey ones to meet different requirements. The underground commercial street connects with the civic eco park, creating a three-dimensional commercial space which is more functional. The buildings are designed in simple European style.

项目概况

项目总用地面积8万 m^2，位于泉州市西北部，晋江西岸，拥有一线江景，户型包括高层、中高层、多层，满足人们各种心理需求。地下商业街和市政生态公园的连接，构成三维立体商业空间，提高景观观赏性和实用性，建筑为简约欧式风格。

Landscape Design Idea

The purpose is to create a high-quality, natural and ecological, and exotic-style environment. The 50 m long green belt serving as the demonstration area of the project, is designed with landscape of Pan Southeast Asian style (Bali style), trying to present the luxury of a five-star hotel. Bali style combines with modern ideas, making the demonstration area full of the exotic flavors and the modern elegance.The design highlights the integration of the natural environment and the Bali style, and the complementary design concept between the natural elements and luxurious architectures. It adopts the tropical style of the paradise resort and simplicity and comfort of Bali Islands to create a Garden of Eden in the city which features exotic style and cozy holiday atmosphere.

景观设计理念

景观设计理念为打造五星品质，凸显尊贵大气，崇尚自然生态，塑造异域风情。50 m 绿化带作为楼盘形象展示区，景观风格定位为泛东南亚（巴厘岛风情），力求打造五星酒店式的奢华品质，融入巴厘岛风情与现代简约理念，使整个展示区既有异域风情品质感，又具简约时尚感。设计体现天然环境与社区巴厘岛风情的互融、自然元素和豪华构筑互补的设计理念，运用巴厘岛度假天堂的热带风情及巴厘岛的质朴闲适，营造出包含异域风情、惬意高雅的度假气氛，打造城市中的梦幻伊甸园。

Site Plan 总平面图

Landscape Design

The design for the demonstration area has taken advantage of the natural environment to match its natural background (Jinjiang River), and at the same time, to guarantee the privacy of the community. The landscapes has followed the style of the demonstration area to be simple, natural and ecological, creating a garden-style living environment. The designers have made full use of the topography, hard landscape and vegetation to create changeable and quite spaces. In addition with the lush trees, beautiful flowers, luxurious and mysterious artworks, as well as the natural timbers and stones, it has created a green community with exotic cultural atmosphere.

景观设计

展示区充分配合当地的自然环境，与天然晋江背景完美结合，同时宣传保证了社区的私密空间。园区内延续展示区的风情理念，以简约、生态、自然为主题，营造生态花园式的生活环境。空间上讲究曲径通幽，利用丰富的地形、硬景、植被营造多变的空间；讲究自然、生态，茂林丛生，花鸟争鸣，使社区沉静在绿意盎然的环境中；讲究精致、奢华、神秘，精巧艺术化小品布置在园区每个角落，采用天然的木材、石材，使社区更具异域文化色彩。

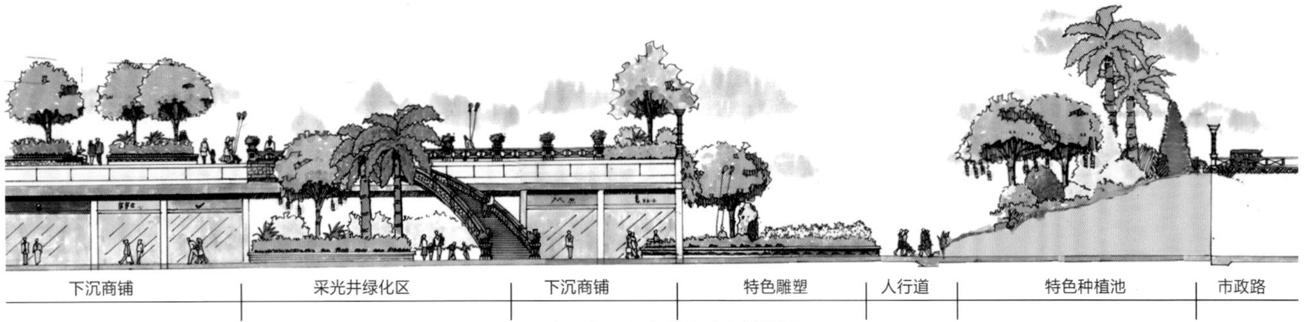

Section A-A1 A-A1 剖面图

Section B-B1 B-B1 剖面图

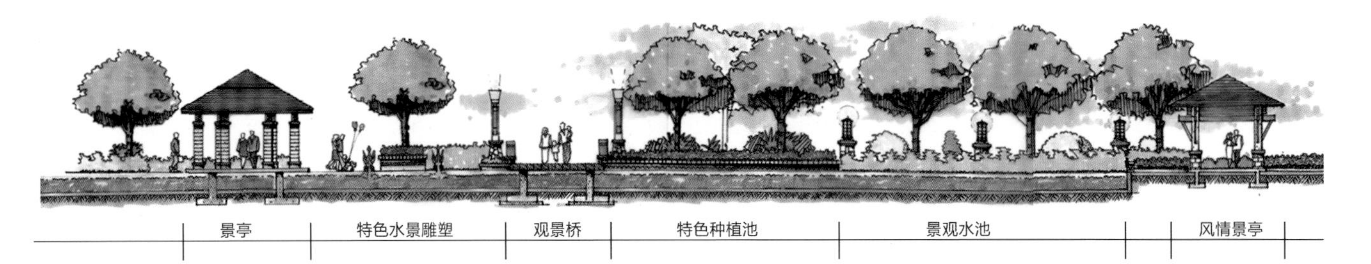

South District in Park Center Garden Section 园内南区中心庭园剖面图

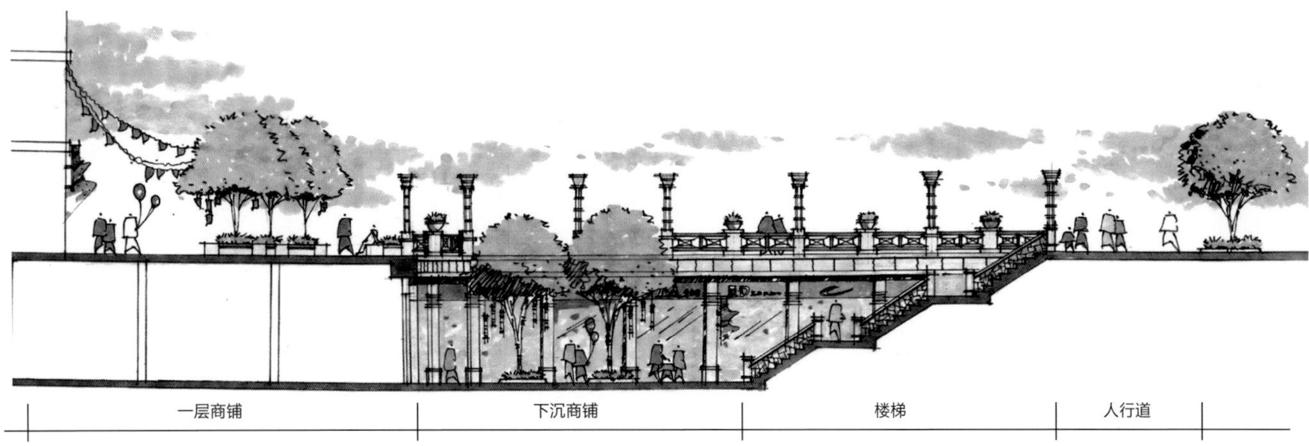

Sinking Commercial Street Section 下沉商业街剖面图

KEYWORDS 关键词

Tropical Scenery
热带风光

Sculptures
雕塑小品

Water Features
大面积水景

Southeast Asian Style
东南亚风格

Location: Nanning, the Guangxi Zhuang Autonomous Region
Landscape Design: KAS Design Group
Land Area: 47, 618.93 m^2

项目地点：广西壮族自治区南宁市
景观设计：深圳凯斯筑景设计有限公司
占地面积：47 618.93 m^2

Nanning Ronghe Central Park
南宁荣和中央公园

FEATURES 项目亮点

To maintain the natural feature is the focus, making full use of the existing materials to accentuate a simple but cozy holiday style: a simple, exquisite, relaxed as well as fashionable lifestyle.

项目还原最自然的风情，充分运用当地材料，强调简朴、舒适的度假风情，营造简洁、精巧、轻松、时尚的生活方式。

Overview

Nanning Ronghe Central Park is located in No. 29-1, Dongge Road, Qingxiu District, Nanning City (former location of party school). The project is aimed to create a green and mixed city close to Nanning River in its center district.

项目概况

荣和中央公园位于南宁市青秀东葛路29-1号（原区党校校址）。项目立足于打造了一个都市中心区的绿色之城、融合之都、邕水之邑。

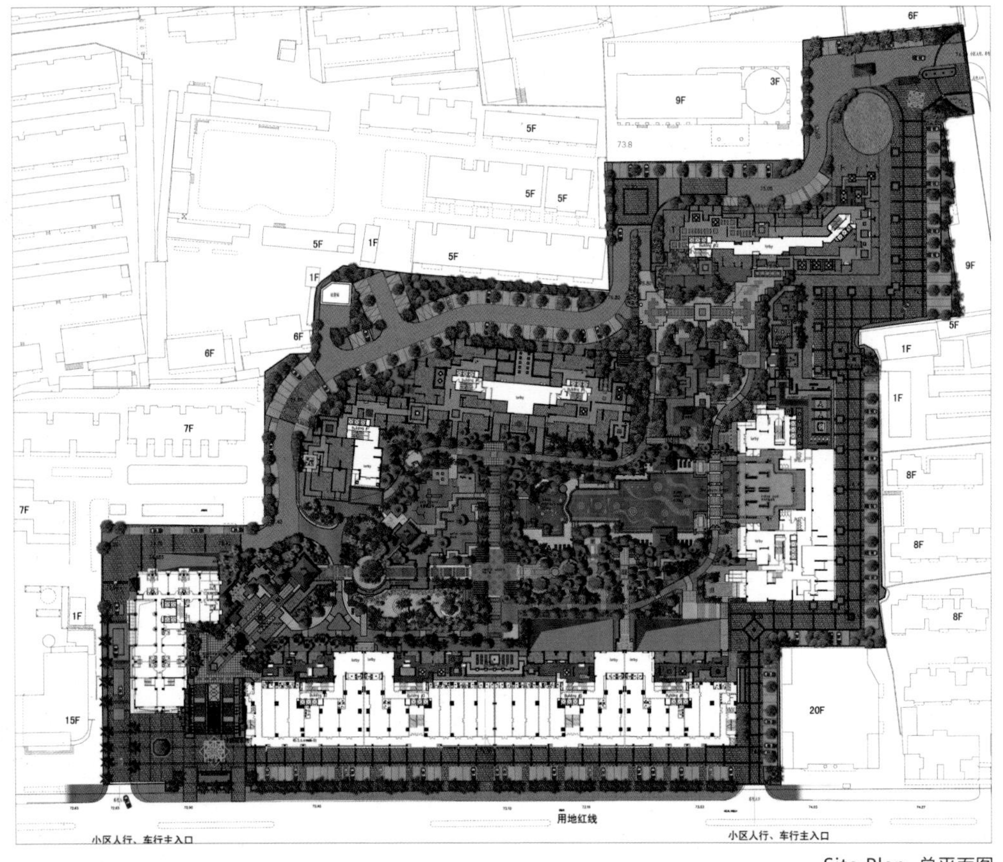

Site Plan 总平面图

Design Objective

Nanning is the permanent site of CAEXPO. KAS whereby designed a landscape garden of Southeast Asia style — Nanning Ronghe Central Park. The foreign elements present distinctive features of this project to an extent, enrich owners' lifestyle and add more colors to the architectural culture of Nanning City.

设计目标

南宁成为东盟博览会的永久会址，设计单位借此打造出荣和中央公园——东南亚风情景观园林。这些异国元素既在一定程度上体现楼盘特色，丰富了业主的生活，较容易为购房者认同、接受，更为南宁的城市建筑文化涂上一抹鲜艳色彩。

Landscape Planning

People can have a good time in the landscape style of this residential area, enjoy the tropical scenery and experience the strong national culture integrated with western customs. To maintain the natural feature is the focus, making full use of the existing materials to accentuate a simple but cozy holiday style: a simple, exquisite, relaxed as well as fashionable lifestyle. Its unique style is reflected by the following factors:

First, as by the concept, showing respect to nature the design creates leisure and a comfortable lifestyle for the client and arranges public spaces for rest and entertainment. Each client is given consideration for their own lifestyle.

Second, we mainly use natural stones and woods as materials for landscape elements to show a natural and artistic beauty of Southeast Asia style. The pavilion, wood paving and stone sculpture are the popular landscape elements in Southeast Asia garden.

Third, we use a variety of water features like pool water, cascade and fountain, to increase the vitality and beauty with distinctive sounds, shapes and colors. Water feature which is the most typical element in Southeast Asia's landscape usually will be decorated with various landscaping features such as trellis, waterside pavilions, pedestrian bridges and step stones along its edge to present an area with heavy Southeast Asia style.

Fourth, Southeast Asia is home to art so that application of rough and original art elements such as sculptures, feature walls, paving, pottery pots and other local features is commonly seen in landscape designs. They are distinctive visual symbols of Southeast Asia style and are carefully scattered throughout the area, some of which seem to be colorful "skins" for the floor or to be pleasing ornamental products like sculptures.

景观规划

本小区景观风格既有着极富特色的热带风光、浓郁的民族文化，又融合了西方的生活习惯、文化理念。它最大特点是还原最自然的风情，充分运用当地材料，强调简朴、舒适的度假风情，提倡简洁、精巧、轻松、时尚的生活方式。它的独特风格主要体现在以下方面：

第一，理念上，尊崇自然，致力于为业主创造轻松、闲适的度假生活。在其景观设计中，会考虑到各个年龄层次业主活动半径的差异，为他们分别设置了休憩、游乐的公共空间，保证各个年龄层都能在此享受度假生活。

第二，景观选材上，以天然原生态的石、木为主，表现出东南亚自然、粗犷的艺术美，木亭、木质铺装、石头雕塑等景观元素在东南亚景观中都比较常见。

第三，应用多种水的形式营造大面积水景，如池水、跌水、喷泉等，通过其各具特色的声音、造型、色彩等，为环境增添生机和美感。水景是东南亚景观中最具特色的元素，在其沿线，往往会穿插点缀各种景观小节点，如喷水兽、花廊、亲水木亭、小桥、汀步等，是东南亚风格体现最为充分、气氛最为浓郁的区域。

第四，粗犷、原生的艺术元素应用，东南亚是艺术之乡，富有当地特色的雕塑小品、景墙、铺装、陶罐等在景观风格中都有所体现，它们或作为铺装散落在小区各地，充当地面色彩各异的“皮肤”，或作为雕塑等观赏品娱人耳目，它们是东南亚风情最为鲜明的视觉符号。

Mediterranean Style

地中海风格

Handicraft
手工艺感

Slope Landscape
坡地造景

Courtyard Spirit
庭院精神

KEYWORDS 关键词

Oceanic Culture
海域文化

Distinctive Features
特色鲜明

Mediterranean Style
地中海风情

Mediterranean Style
地中海风格

Location: Zhuzhou, Hunan
Landscape Design: Guangzhou Bosi Landscape Design Co., Ltd.
Land Area: 68,535.72 m^2

项目地点：湖南省株洲市
景观设计：广州柏思园林景观设计有限公司
占地面积：68 535.72 m^2

Zhuzhou Riverfront Garden
株洲滨江花园

FEATURES 项目亮点

The project adopts the elements with rich Mediterranean styles such as the pavilion, flower corridor, stylish ornaments, stylish pavements, ocean animal ornaments or sculptures that show the "oceanic culture" and scenes with distinctive features to create the thematic landscape and to build a resort-style courtyard space for leisure life.

项目采用富有地中海度假风情的构件，包括亭、花廊、风情小品、风情铺装、体现“海域文化”的海洋动物小品或雕塑等极具有风格特色的场景来塑造主题景观，创造一种度假式的休闲生活庭园空间。

Overview

With a gross land area of 68,535.72 m^2, the project is by the river of Xiangjiang and near the Shifeng Bridge, and it enjoys the unique geographical advantages and the valuable mountain resources. With mountains on its back and facing the water, it's a hard-to-get livable place.

项目概况

项目占地68 535.72 m^2，位于湘江河畔，临石峰大桥，有着得天独厚的地理优势和宝贵的山体资源。靠山面水，占尽地理优势，是不可多得的宜居福地。

Design Objective

The development conforms to the natural conditions and the characteristics of the building such as the linear form, aiming to build a livable landscape with the theme of "oceanic culture" and rich in Mediterranean and resort styles.

设计目标

项目顺应自然条件与建筑的流线造型等特点，意将项目打造成一个富有地中海度假风情的“海域文化”主题宜居景观。

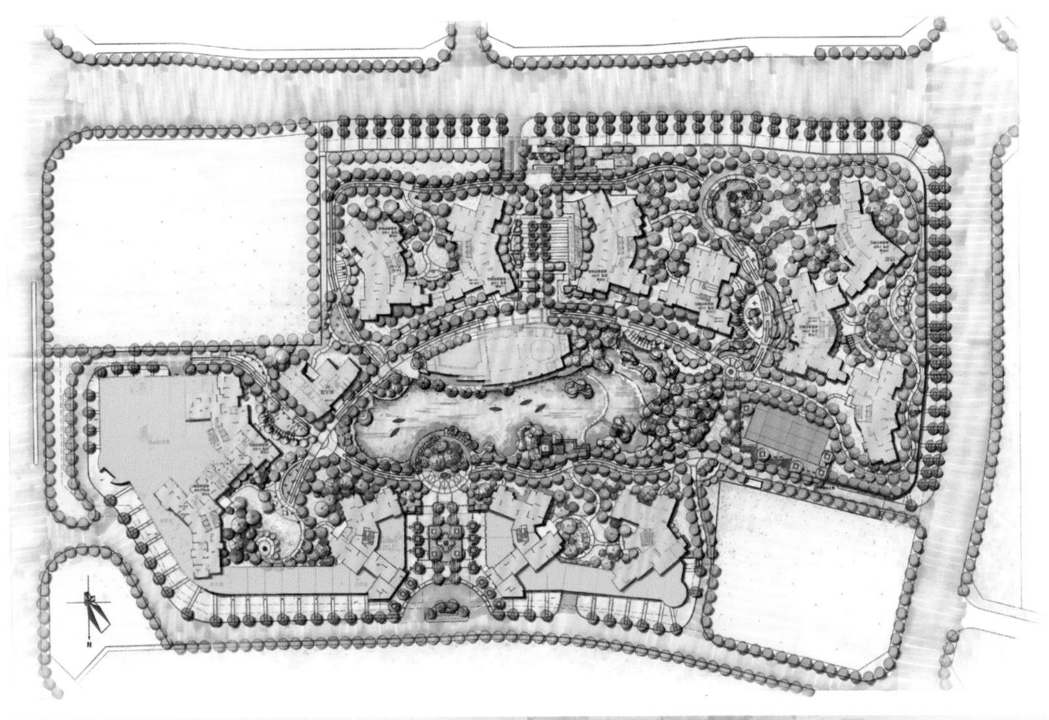

Landscape Planning

The designers have designed various groups of featured scenes and enhanced the effect of Mediterranean style, which includes the commercial street, main entrance of the community, the rain yard and the green yard; the large-scale lake area is arranged with Mediterranean style square, large-scale dropping water landscape, club recreation platform, central pavilion, lakeside pavilion, stepping stones with dropping water, rich lakeside landscapes and so on.

The entire community is with reasonable arrangement, clear layer and distinctive features, and the space is static and dynamic, and is with good opening and closing and clear lines, as well as rich and explicit functions. Wonderful scenery is everywhere, and it changes while one moves, and it's with rich seasonal changes. Beautiful and attractive sceneries can be easily found at the lakeside, under the tree or by the creek and so on.

The project adopts the elements with rich Mediterranean styles such as the pavilion, flower corridor, stylish ornaments, stylish pavements, ocean animal ornaments or sculptures that show the "oceanic culture" and scenes with distinctive features to create the thematic landscape and to build a resort-style courtyard space for leisure life.

景观规划

项目设计的多组具有风格特色的场景，加强地中海风格效果，包括商业街、园区主入口、听雨小苑、绿林小苑；中心有大型湖区设有地中海风情广场、大型跌水景观、会所休闲平台、湖心水榭、湖边亭、水上跌水汀步、丰富的湖岸景观等。

整个园区中，布局合理、主次分明、风格特色鲜明，空间有动有静、开合有致、流线清晰、功能明确丰富。处处佳景、步移景移、季相丰富，无论在湖边、树下、小溪旁，每一处都可以找到让人为之心动的景色。

项目采用富有地中海度假风情的构件，包括亭、花廊、风情小品、风情铺装、体现“海域文化”的海洋动物小品或雕塑等极具有风格特色的场景来塑造主题景观，创造一种度假式的休闲生活庭园空间。

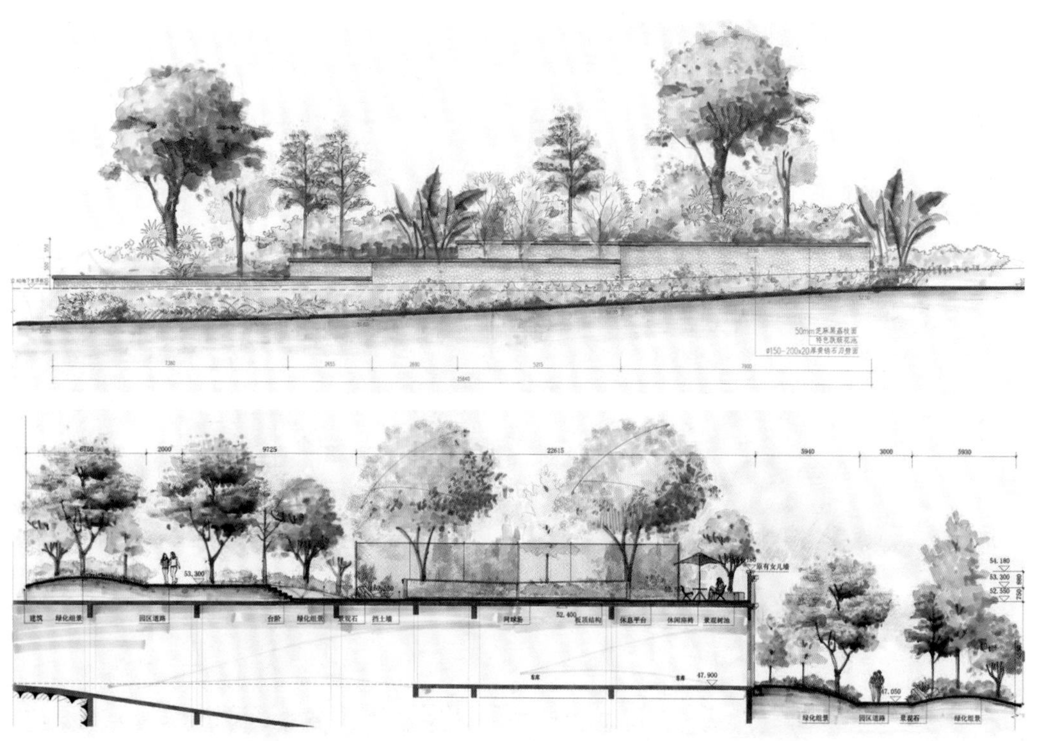
2000
9725
22615
3940
3000
3930
53.300
原有女儿墙
54.180
53.300
52.550
建筑
绿化组景
园区道路
台阶
绿化组景
景观石
挡土墙
网球场
52.400
板顶结构
休息平台
休闲座椅
景观树池
车库
车库
47.900
47.050
绿化组景
园区道路
景观石
绿化组景

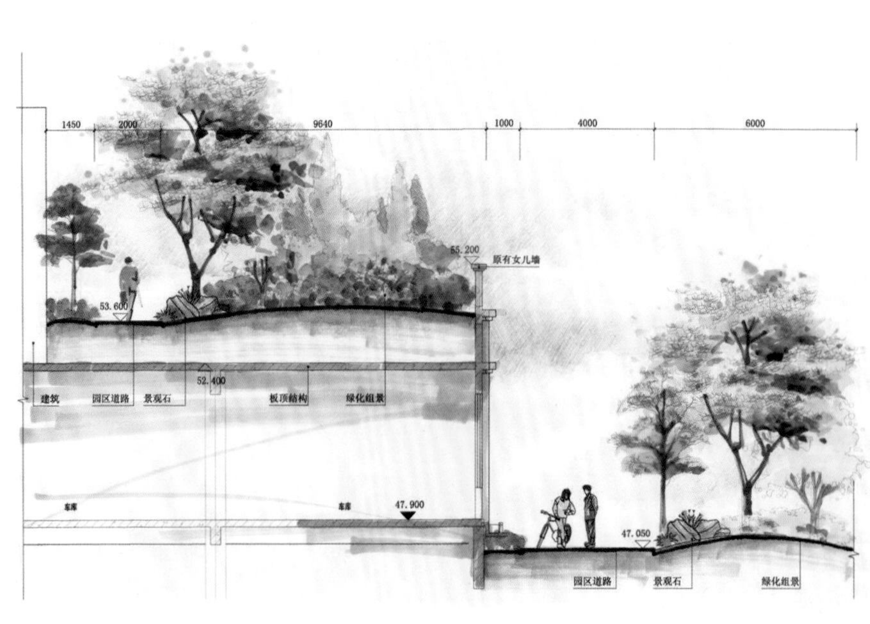
1450
2000
9640
1000
4000
6000
55.200
原有女儿墙
53.600
52.400
建筑
园区道路
景观石
板顶结构
绿化组景
车库
车库
47.900
47.050
园区道路
景观石
绿化组景

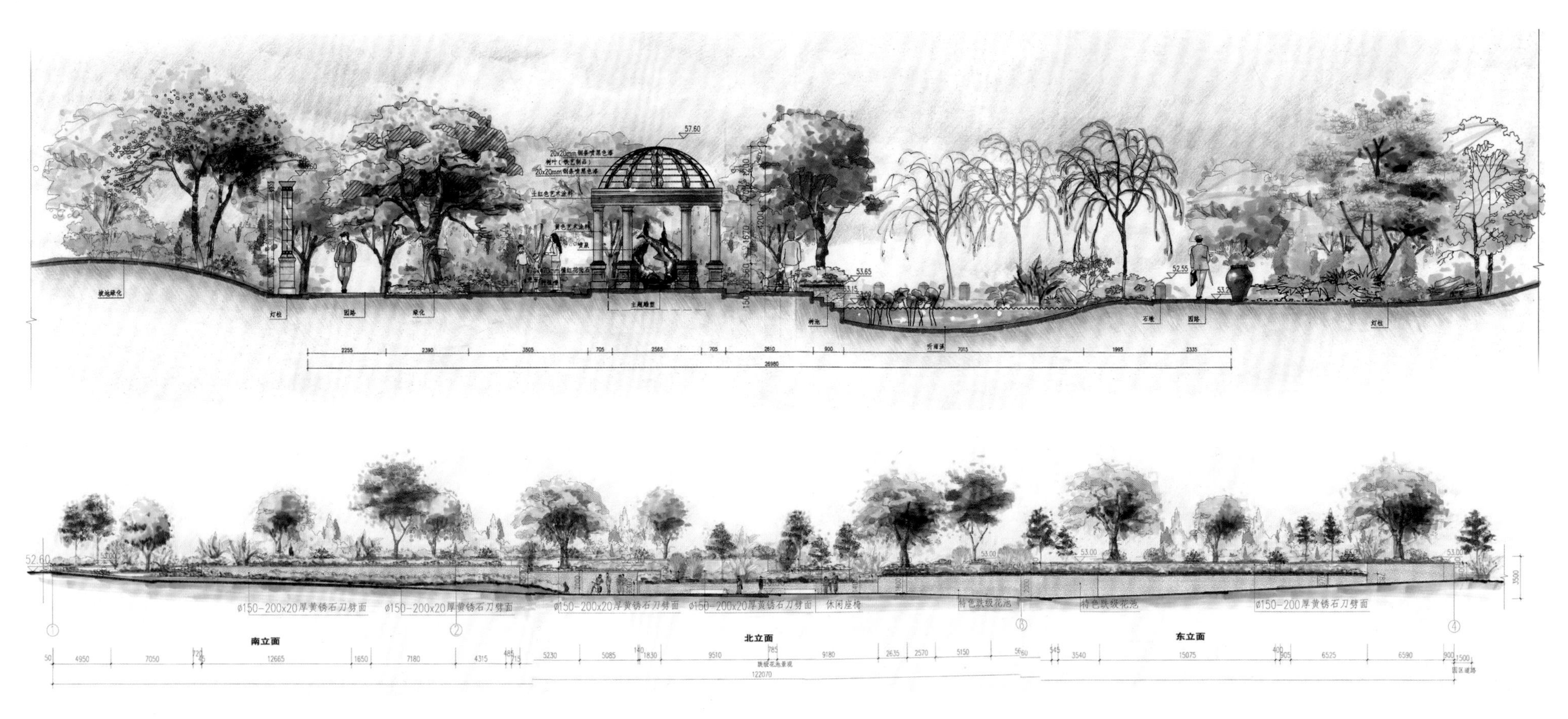
57.60
坡地绿化
灯柱
园路
绿化
主题雕塑
53.65
53.15
树池
52.55
石墩
园路
灯柱
2255
2390
3505
705
2565
705
2610
900
7015
1995
2335
26980
52.60
ø150-200x20厚黄锈石刀劈面
ø150-200x20厚黄锈石刀劈面
ø150-200x20厚黄锈石刀劈面
ø150-200x20厚黄锈石刀劈面
休闲座椅
特色跌级花池
特色跌级花池
ø150-200厚黄锈石刀劈面
53.00
53.00
53.00
3500
南立面
北立面
东立面
50
4950
7050
720
12665
1650
7180
4315
5230
5085
1830
9510
785
9180
2635
2570
5150
3540
15075
6525
6590
900
1500
跌级花池景观
122070
园区道路

1、2栋
物业服务中心

KEYWORDS 关键词

Jungle Fun
丛林野趣

Slope Landscape
坡地造景

Five-layer Landscape
五重绿化

Mediterranean Style
地中海风格

Location: Songjiang District, Shanghai
Landscape Design: L&A Design Group
Landscape Area: 92,734 m²

项目地点：上海市松江区
景观设计：奥雅设计集团
景观面积：92 734 m²

Shanghai Longfor Affecting Yard
上海龙湖好望山

FEATURES 项目亮点

Designers put the highlight on gorgeous and meandering river, multi-layers plants and thick riverbanks, to create a wild garden that full of jungle fun.

景观设计重点打造了形态唯美、蜿蜒曲折的溪流，层次丰富的植被，繁绕茂密的水岸线，使人们仿佛置身野外花园，充满丛林野趣的氛围。

Overview

Nestled in Songjiang New Town, Shanghai, at the east of Renmin North Road, west of Longma Road, north of Meijiabang Road and south of Guangfulin Road, close to Songjiang University Town Station of Line 9, the project has the volume of 148,000 m², and 186 courtyard villas, 3 decorated high-rise buildings and 5 large-sized apartments. While courtyard villas are the original products of Longfor, to represent the concept of works full of dynamic.

项目概况

龙湖好望山坐落于松江新城板块，西至人民北路，东至龙马路，南至梅家浜路，北至广富林路，毗邻轨交 9 号线松江大学城站。项目总面积达 14.8 万 m²，规划为 186 套合院别墅、3 栋精装小高层、5 栋大平层公寓。其中，合院别墅为龙湖独创，展现“生命洋溢的作品”的理念。

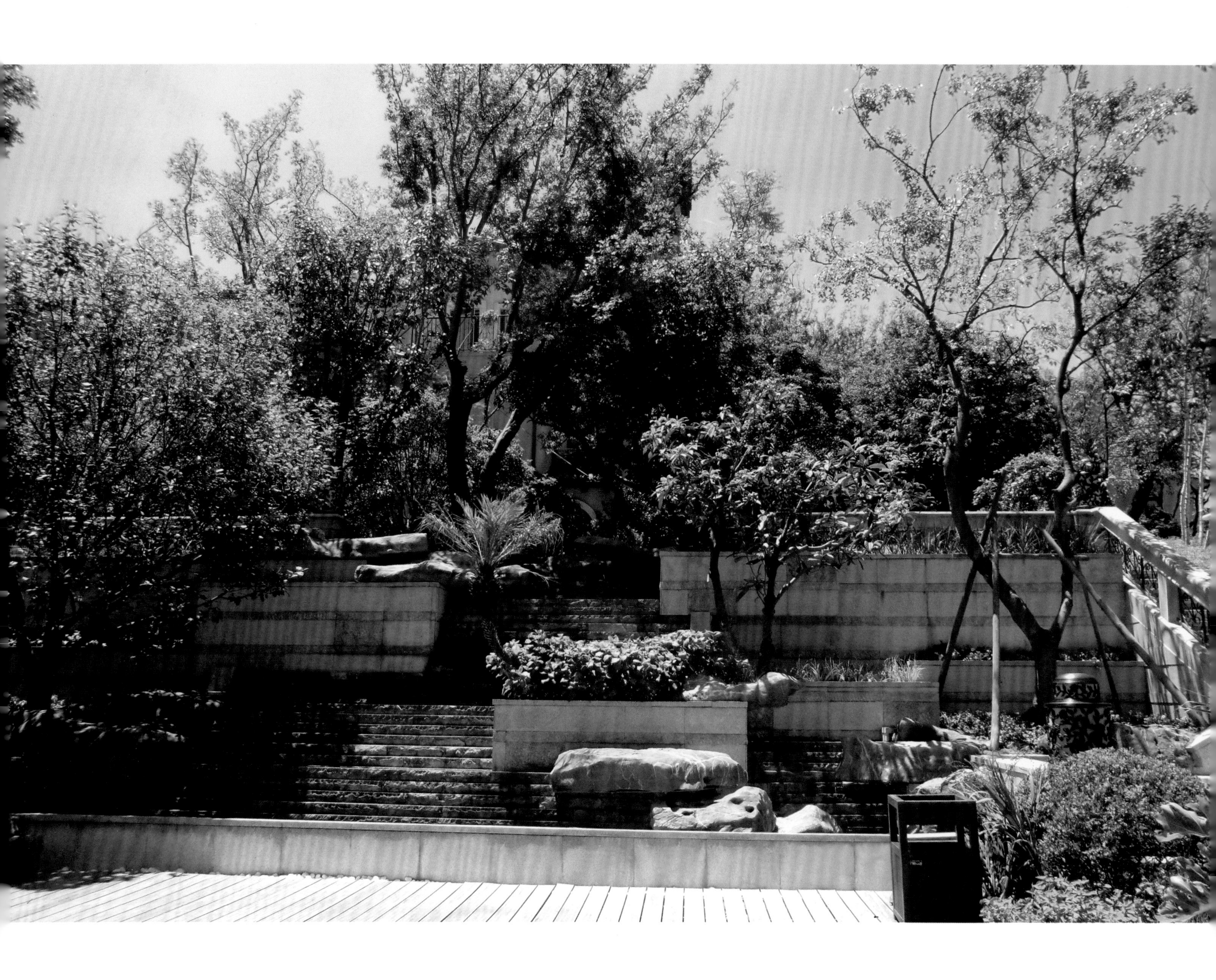

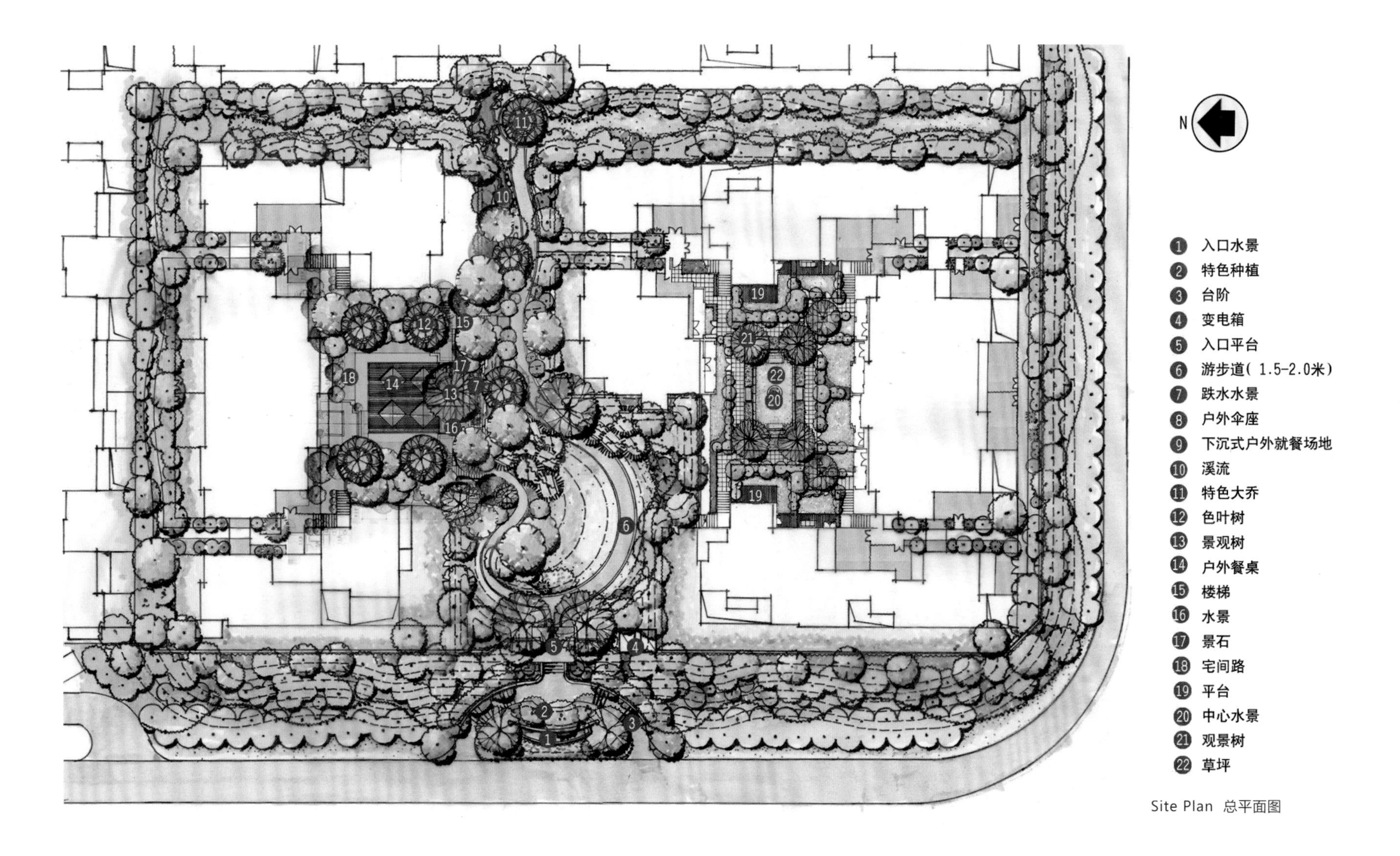

Site Plan 总平面图

Design Objective

Adopting Mediterranean Style for landscape, designers choose slope and double first layer for landscape and upgrade original Longfor five-layer landscape, to achieve ecological landscape experience.

设计目标

项目的景观风格定位为地中海风格。项目整体采用坡地造景、双首层设计，并在龙湖五重绿化的基础上进行升级，达到全生态系统景观体验。

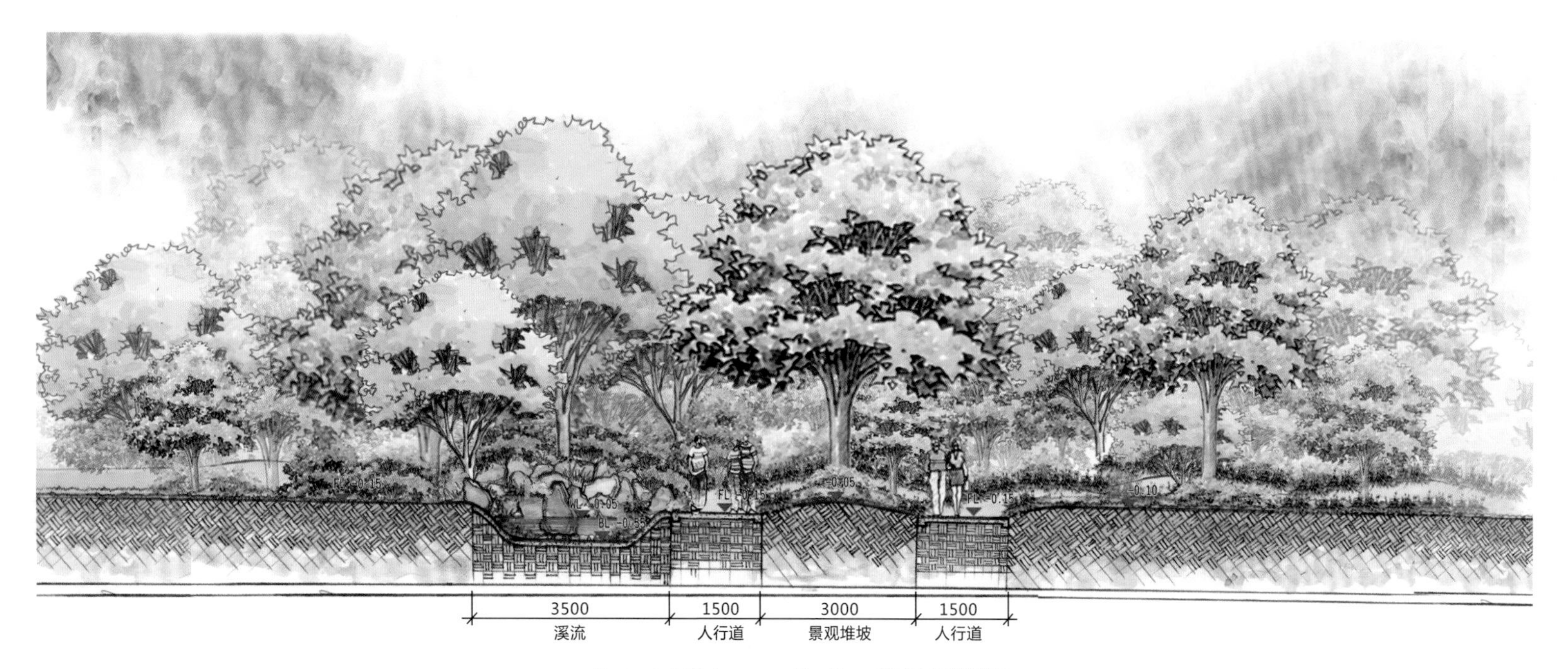

Endpoint Waterscape Section 端点水景剖面图

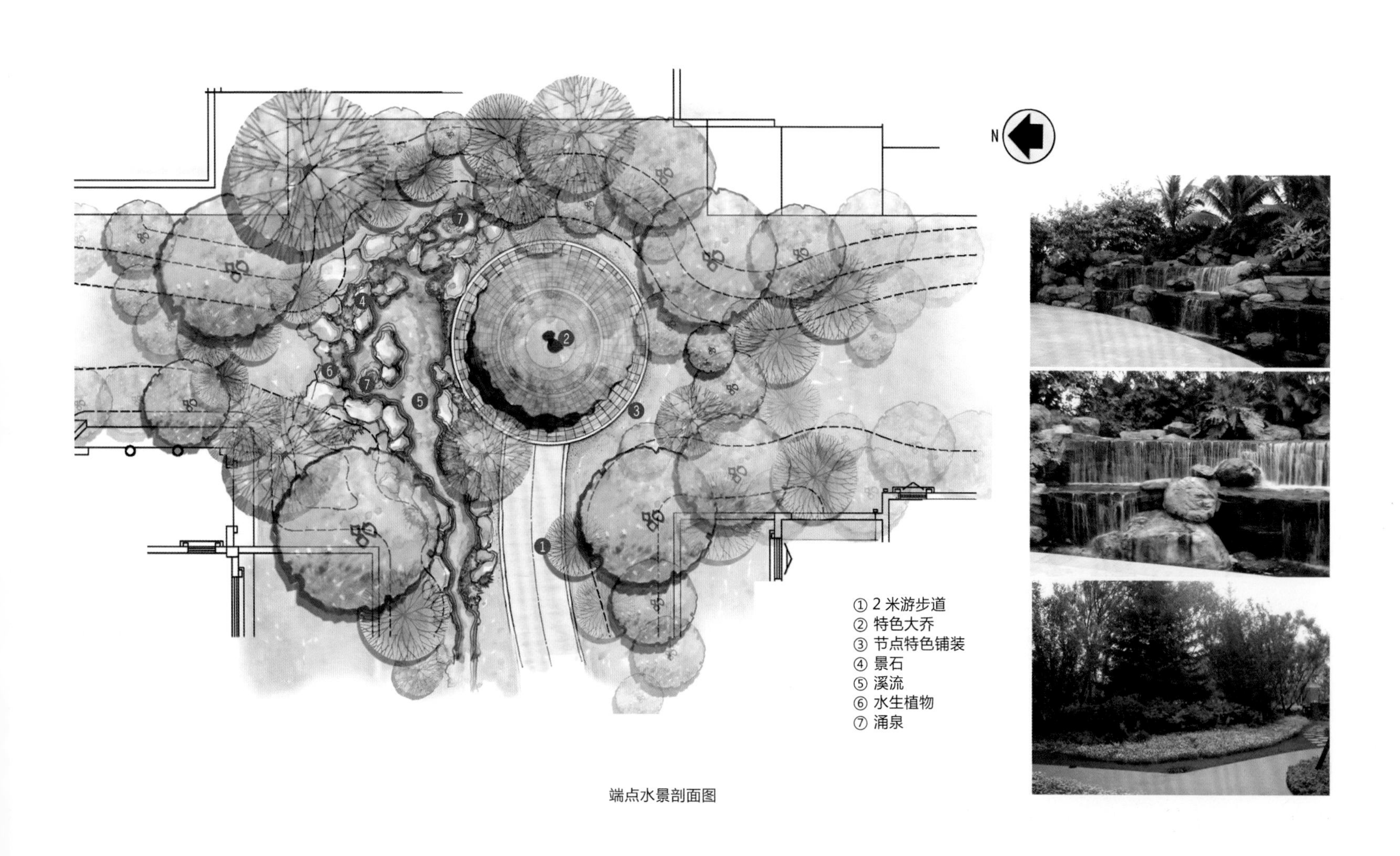

端点水景剖面图

Landscape Planning

Employing naturalism as the theme, designers put the highlight on gorgeous and meandering river, multi-layers plants and thick riverbanks, to create a wild garden that full of jungle fun.

The sales center and sample courtyard are formerly completed former, cooperating with unique courtyard architecture and focusing on plants design to create a natural sunshine yard with rivers around. Stepping into the garden from main entrance, a grand grass comes into eyes with flourishing plants around to form independent space and exalted feelings in the same time. Meandering rivers guide residents into the quiet pathways to make full use of masking characteristic of plants in the limited space. The openness and closeness have created colorful landscape layers and visual rhythm.

景观规划

项目以自然主义为主题，景观设计重点打造了形态唯美、蜿蜒曲折的溪流，层次丰富的植被，繁绕茂密的水岸线，使人们仿佛置身野外花园，充满丛林野趣的氛围。

先期完成的部分是售楼处及样板庭院区，结合独特的合院建筑，以植物设计为重点，旨在打造一个自然、水系环抱的阳光花园。从主入口进入，是一片开阔的草坪，周围用繁茂的植物围合，有效塑造独立空间的同时，又突出尊贵的感觉。蜿蜒的溪流，引导行人的视线进入幽静的小径，充分利用有限的空间将植物的遮蔽性发挥到最大。一开一闭，创造了丰富的景观层级及视觉节奏。

KEYWORDS 关键词

Stylish Square
风情广场

Landscape Groups
组团景观

Characteristic Decorations
特色小品

Mediterranean Style
地中海风格

Location: Foshan, Guangdong
Developer: Kaisa Group Holdings Ltd.
Landscape Design: CSC Landscape
Landscape Area: Approx 140,000 m^2

项目地点：广东省佛山市
开发商：香港佳兆业集团
景观设计：深圳市赛瑞景观工程设计有限公司
景观面积：约 14 万 m^2

Kaisa Golden World, Ronggui
佳兆业容桂金域天下

FEATURES 项目亮点

In consideration of the cultural and natural characteristics of the site, the designers have combined the Spanish royal garden elements with Lingnan garden style to extend along the cultural and eco axis and penetrate into the building groups through urban commerce interface.

项目考虑到基地城市人文要素与自然要素特征，设计采用西班牙皇家园林元素与岭南园林风格的对话方式，通过城市商业延展界面，由中心景观人文生态轴向各组团空间渗透。

Background

Located in the central eco residential area of East Ronggui, Shunde District of Foshan City, the project is next to the shopping center and shopping mall. Kaisa is dedicated to create a high-end landmark residential community of Spanish style that is independent from the noisy surroundings. It will pay more attention to shaping the living atmosphere of the Spanish garden style and bringing the friendliness to the green residential groups. The commission includes the landscape design for the phase one's clubhouse, the swimming pool area and the waterfront landscape area.

项目背景

容桂金域天下项目位于佛山市顺德区容桂东部中心生态住宅区，紧邻大型购物中心和大型商场。佳兆业致力将其打造成为独立于商圈之外、闹中取静的西班牙滨水地标性高档住宅小区，项目将会更强调西班牙园林式生活氛围和组团绿化的亲切感。金域花园设计范围为一期会所和泳池展示区及滨水景观区的综合景观设计。

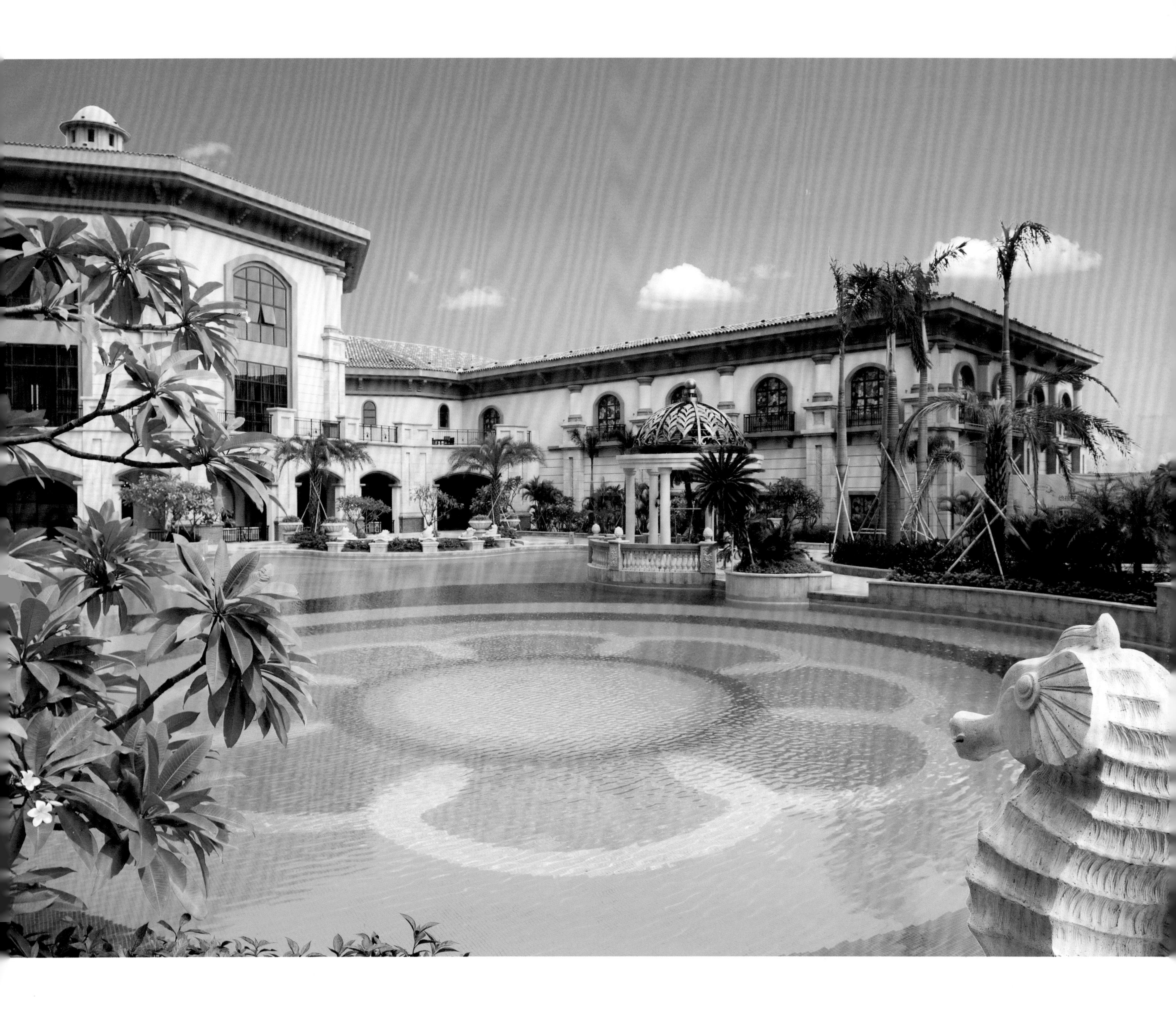

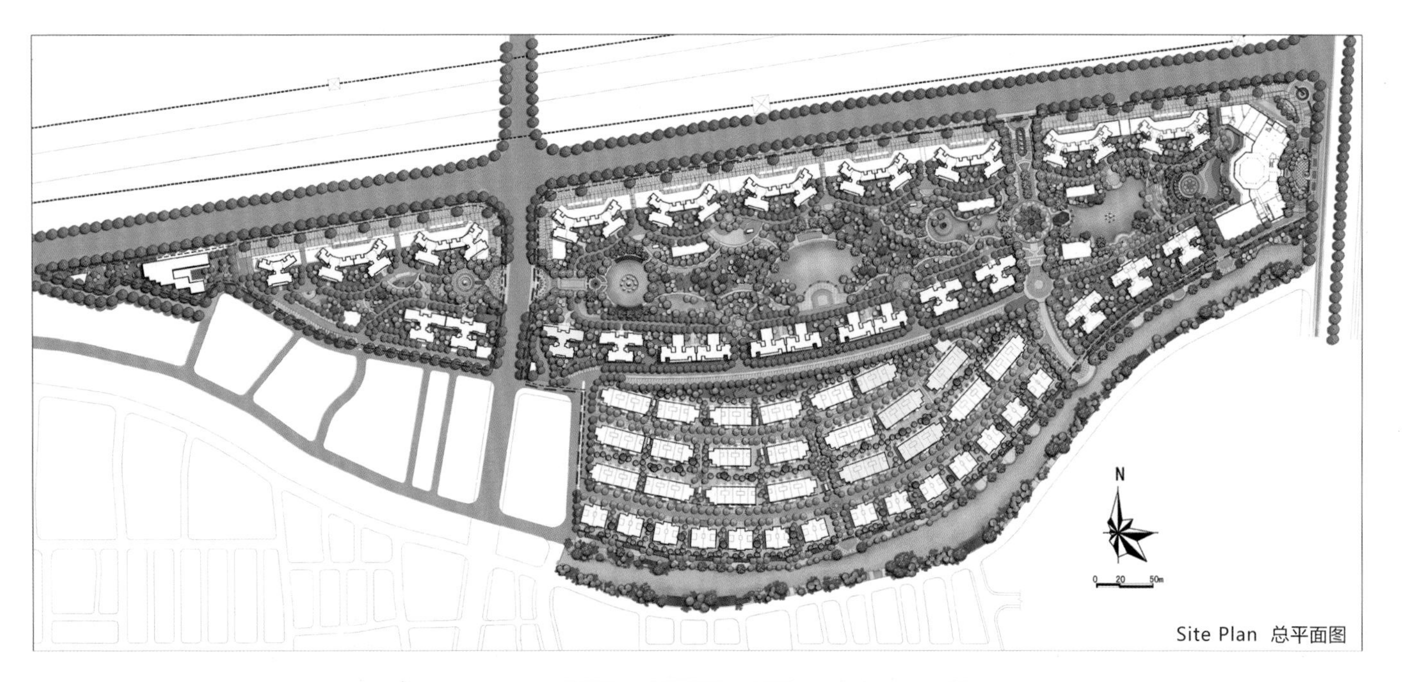

Site Plan 总平面图

Design Idea

The project is designed with the idea of "The Resort Life Style in Hidden Valley". "Resort Life Style" refers to a kind of postmodern lifestyle that is very popular in today's Europe and America. The original meaning is to travel to a tourist attraction, but now it is extended as a kind of high-end and leisurely lifestyle. "Valley" is a natural structure that combines lawns, streams and other natural elements, which represents a kind of natural and ecological living environment. "Hidden" has two meanings: firstly, after experiencing the vanity, people will turn back to nature, so "Hidden" shows the up-class's living requirements; secondly, the buildings hidden in green show a kind of low-key luxury and they will warm people like the sunshine with their green roofs interfacing with the blue sky and the green trees.

The urban context of a city is composed by multi elements, natural or cultural. The garden of the Golden World has shown the natural context of the city and created a world of peace that is hidden behind the hustle and bustle. Nature, art and leisurely lifestyle are what the up-class want to enjoy and what the designers want to create. Therefore, the Golden World's garden has integrated all the natural elements, including hill stones, waters, sunshine and even air. At the same time, it features the artistic quality which allows people to enjoy the "Resort Life Style" in the "Hidden Valley".

设计理念

项目以“The Resort Life Style in Hidden Valley”为设计理念。“Resort Life Style”是一种现代欧美流行的后现代生活方式，原义为前往圣地旅游度假，现引申为高尚的自然休闲生活方式。“Valley”在自然生态概念中代表了草地、溪水等自然要素的结合体，而在生活中则代表了自然生态的生活场所。“Hidden”意为隐藏的：其一，社会顶级圈层浮华表面的炫耀之后，便是低调、内敛的回归，因此“Hidden”表现了社会顶级圈层的居住所求；其二，掩映在绿荫丛林中的建筑，以其厚重、质朴的形体，蕴藏其低调的奢华，红色的屋面与蓝天、绿树相映衬，带给人阳光般的温暖。

每个城市的文脉是由多个点构成的，或是自然或是人文。金域花园所体现的独特气质恰恰是城市的自然脉络，是隐于世的旷达神悠之境。自然、艺术、隐隐的休闲生活是现代都市高尚阶层的诉求，亦是设计师空间设计的理想。它融合了自然的一切元素——山石、水、阳光甚至空气，同时又兼备了艺术品质，让人们在“Hidden Valley”中享受少数人拥有的“Resort Life Style”。

Landscape Design Skills

In consideration of the cultural and natural characteristics of the site, the designers have combined the Spanish royal garden elements with Lingnan garden style to to extend along the cultural and eco axis and penetrate into the building groups through urban commerce interface. The framework is formed by the Spanish-style square, landscape architectures and courtyards. And the landscape avenue, high-rise area, swimming pool and clubhouse, as well as the waterfront landscape area are all designed with the landscape elements and the axial symmetric patterns of the Spanish royal gardens. In the "Hidden Valley", it takes advantages of the undulating topography to show the natural elements of the site. The garden housing group and the driveway are designed with the elements of royal gardens to enable people to experience the artistic atmosphere of the Spanish-style community.

Taking advantages of the topography, the designers have combined the outdoor landscape planning with the living spaces perfectly. The landscape spaces between buildings are well designed by different skills with different plants and Spanish-style elements to highlight their unique identities and create multi-level and diversified space effect. From south to north, a landscape avenue runs through the community to connect the high rises, swimming pool and clubhouse, villas and garden houses, and the riverside landscape belt.

The swimming pool is connected with the unique cascade and the landscape lake. The turf slope extending into water looks interesting. The plants are carefully selected to decorate different landscape themes in four seasons. Different colors, classical patterns, pavements and materials are combined with green items to create a natural, ecological, beautiful and comfortable environment for living. The landscape design for each building group tries to provide every house with a green world in this hustling and bustling city.

景观设计手法

考虑到基地城市人文要素与自然要素特征，设计采用西班牙皇家园林元素与岭南园林风格的对话方式，通过城市商业延展界面，由中心景观人文生态轴向各组团空间渗透。以西班牙风情广场、西班牙小品、西班牙庭院等景观元素为立足点进行设计构思。景观大道、高层区、泳池会所区、滨水景观区引荐了西班牙皇家园林景观空间及代表性元素符号，以轴线对称的方式体现了城市人文要素；溪谷自然景观带则利用地形的穿插变化体现基地园林景观的自然要素；洋房组团和车道则以自然为基调，点缀皇家园林的元素，让人们仿佛置身于具有艺术风范的西班牙社区。

在细节处理上设计师结合地域特征将户外的景观规划与居住空间完美结合，组团景观带配备西班牙皇家园林及西班牙特色小品，运用不同的造景手法和四季植物搭配，来突出每个组团建筑的个性特征，营造出层次丰富、多意境的空间景观效果。在整个区域景观中，由一条景观大道贯穿小区南北，将高层区、会所泳池区、别墅洋房区以及滨河景观带关联为一个整体。

泳池连接着特色跌水和景观湖，将草坡以自然曲线入水，充满自然情趣。根据季节精心挑选的四季植被等丰富了主题组团景观，结合色彩、古典的图案、铺地、材料等元素，缀以绿色小品，布局自然、错落有致，营造自然、生态、优美、宜居的居住区环境。社区内部组团式景观设计尽量将风景呈现给每一个窗台，让每一户居民都可以在都市喧嚣中有属于自己的绿色方舟。

Art Deco Style

ART DECO 风格

Geometrical Volume
几何造型

Curved Lines
富有曲线

Harmony between Human and Landscape
人景交融

KEYWORDS 关键词

Decorative Style
装饰风格

Courtyards Landscape
中庭景观

Connected by Water Features
水系连接

Art Deco Style
ART DECO 风格

Location: Gaoxin District, Chengdu, Sichuan
Developer: CITIC (Chengdu) Construction Co., Ltd.
Landscape Design: Chengdu JZFZ Architectural Design Co., Ltd.
Land Area: 150,000 m^2

项目地点：四川省成都市高新区
开发商：成都中信城市建设有限公司
景观设计：成都基准方中建筑设计有限公司
占地面积：15 万 m^2

CITIC Future City
中信未来城

FEATURES 项目亮点

For courtyard of south plot, water features are used to connect each activity space and so that to make the community more dynamic and vivid, by making full use of water resource to extend the water features to the sub-entrance of south plot to improve its landscape quality.

南区中庭以水系连接各个活动场地而赋予社区灵魂，并充分利用水资源，把水系延续到南区次入口，提高入口景观质量。

Overview

Located at Shuxin East Road of Gaoxin District, Chengdu City, the project occupies the core location of Chengdu New Town development, as sitting at north of Modihe Park, south of Longfor development, west of Gaozhuan Road and east of Changqing Road, and connecting with Knowledge Park and Chuangzhi Park.

项目概况

项目基地位于成都高新区蜀信东路，中信・蜀都新城的核心位置，南邻摸底河公园，北接龙湖地块，东接高专路，西邻长清路，处于新城知识公园与创置公园的连接带上。

中信|未来城
（北区）

Design Objective

With the ecological principle of "adopting recyclable and reusable materials", designers choose the design theme of "droplets on the keys" and cooperate with original topography to express the theme to the maximum.

设计目标

项目以“材料循环 + 再利用概念”为生态原则，以“琴键上的水珠”为设计主题，结合基地特点，力求实现水元素的主题性表达。

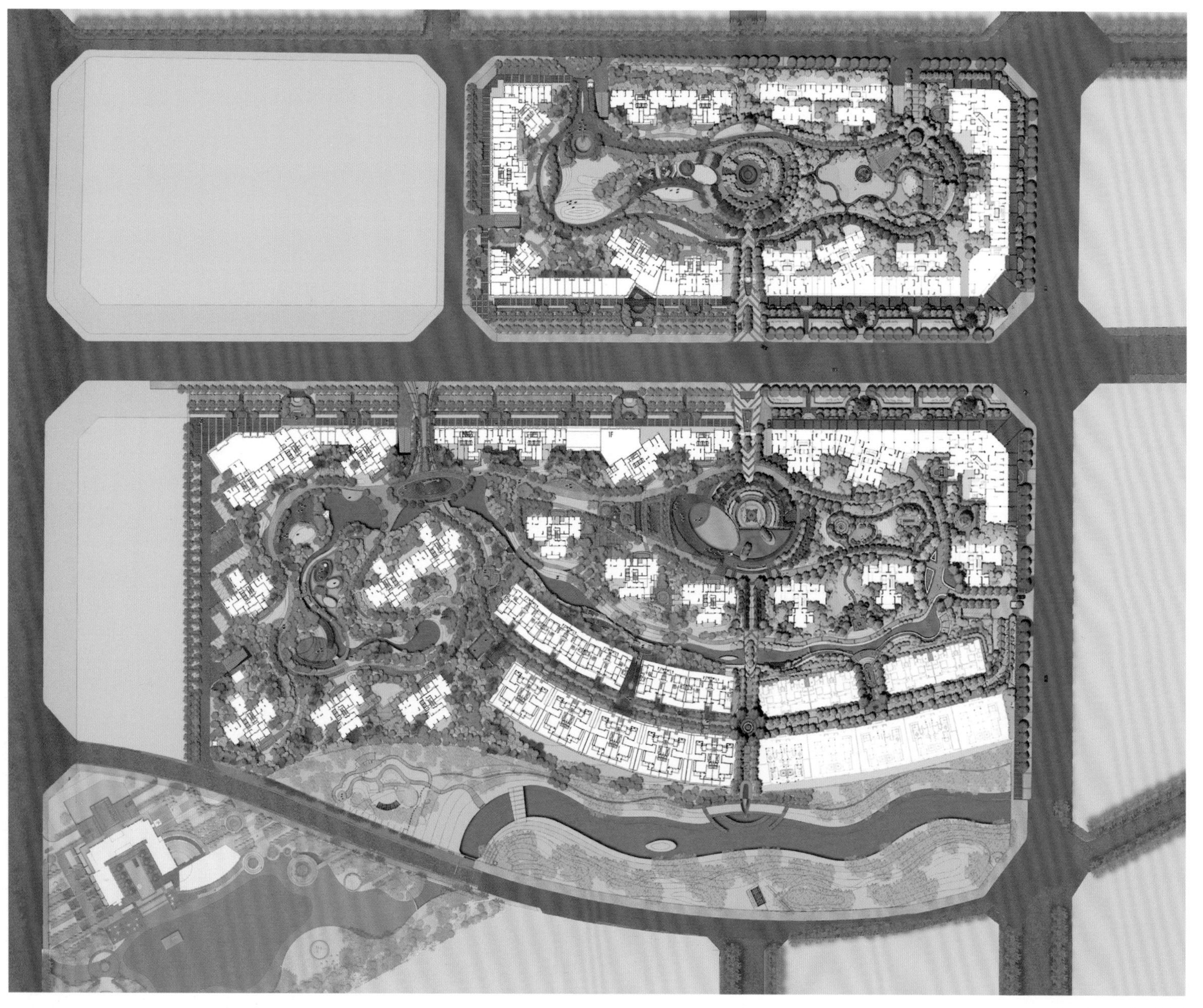

Site Plan 总平面图

Landscape Planning

It is divided into south and north plots with a central axis crossed them and then courtyards landscape are introduced in.

Courtyard of north plot has set a square for activities and some functional spaces to complement the functions of community, to improve residents' degree of satisfaction and landscape viewing.

While for the courtyard of south plot, water features are used to connect each activity space and so that to make the community more dynamic and vivid, by making full use of water resource to extend the water features to the sub-entrance of south plot to improve its landscape quality. Height differences between villas and high-rise residences are softened by water features, in the meanwhile, dignified and unique feelings of independent villas are also being smoothly expressed. Plants are employed at the boundary of landscape and elevated floor to form semi-private spaces as reading and fitness rooms.

景观规划

基地分为南北两区，中轴线贯穿北区和南区，并引入中庭景观。

北区中庭设置适宜社区交流活动广场及一些功能性场地，完善小区功能，提高住户满意度，并使住户拥有良好的景观视点。

南区中庭以水系连接各个活动场地而赋予社区灵魂，并充分利用水资源，把水系延续到南区次入口，提高入口景观质量。利用水系处理洋房区和高层之间的高差关系，并自然体现洋房区的独立尊贵感受。通过植物来过渡架空层与景观的边界，并在架空层形成半私密空间，形成阅读室和健身室。

全城热恋，相约未来城
中信未来城2012年大型交友派对
中信地产
2009-2012年
我们不断升级 持续领先
（七大体系·荣耀升级）

KEYWORDS 关键词

British Garden
英式园林

Dignity & Honor
威仪尊荣

Noble Ambiance
贵族气息

Art Deco Style
ART DECO 风格

Location: Beijing
Landscape Design: United Design Associates
Land Area: 442,600 m^2

项目地点：北京市
景观设计：优地联合（北京）建筑景观设计咨询有限公司
占地面积：442 600 m^2

United Kingdom Palace
永定河 · 孔雀城英国宫

FEATURES 项目亮点

The Victoria garden absorbs the dignity and honor of the European classical garden. The symmetrical axis square, exquisite lawn and gorgeous gardening modeling are full of British noble life ambiance.

维多利亚花园浸润西欧古典园林的威仪与尊荣，对称中轴广场、精美草坪和精美绝伦的园艺造型，洋溢英伦贵族生活气息。

Overview

The United Kingdom Palace is located at the area of the Second Beijing Airport. It's a low plot ratio leisure British garden city. The project has provided the clients with the leisure life of a British small town. The various house types such as garden house, medium high-rise and high-rise combine with the British architectural style, together with the 400 m town streets and the "British garden comprehensively guide-in system" to let the clients feel noble and leisure.

项目概况

永定河·孔雀城英国宫位于北京第二机场区，是一个低容积率的悠享英国花园城。项目为客户提供一种悠享的英国小镇生活。花园洋房、小高层、高层等多样的产品种类，结合典型的英式建筑风格，通过 400 m 长的小镇商街和“英国花园全面导入系统”让业主在项目内感受尊贵与悠然。

UNITED KINGDOM PALACE

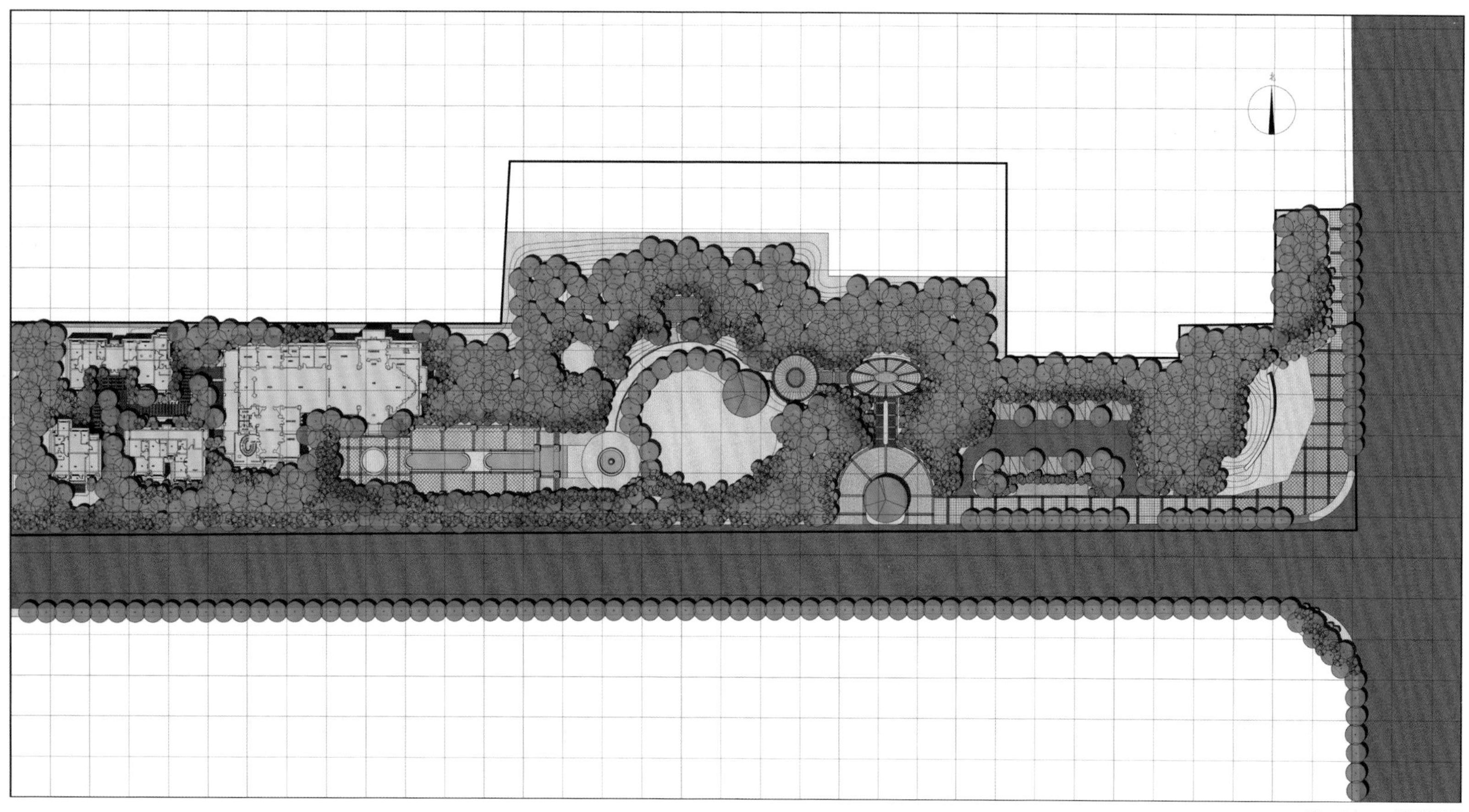

Site Plan 总平面图

Design Objective

The United Kingdom Palace adopts the idea of "each United Kingdom Palace is a British garden". It perfectly combines the Art Deco architectural style of British royal garden, uses the pure British life to inherit the brand connotation of peacock family, inherits the British noble quality, and absorbs the essence of royal palace garden to create an ultimate ideal home.

设计目标

永定河·孔雀城英国宫以"每一个英国宫都是一个英国花园"为理念，将英式皇家园林 ART DECO 建筑风格完美结合，以纯正英伦生活继承孔雀家族品牌内涵，沿袭英伦贵族品质，吸纳皇室宫廷花园精髓，打造理想的终极家园。

Landscape Planning

The landscape design of United Kingdom Palace uses the British palace landscape as design blueprint and strives to combine the elegant, solemn and quaint British landscape with the British gentlemen life, together with the landscapes, buildings, properties, 6,000 m^2 British bilingual kindergarten, the British style business street and the 74,000 m^2 commercial complex, making the United Kingdom Palace an unique British garden city of Beijing area. Besides, the Victoria garden absorbs the dignity and honor of the European classical garden. The symmetrical axis square, exquisite lawn and gorgeous gardening modeling are full of British noble life ambiance.

景观规划

永定河・孔雀城英国宫的景观设计以英国宫廷景观为设计蓝本努力将英式景观的典雅、庄重、古朴与英国的绅士生活相结合，通过景观、建筑、物业以及 6 000 m^2 的英国双语幼儿园，再加上英伦风情商业街和 7.4 万 m^2 的商业综合体，让孔雀城英国宫成为大北京区域独一无二的英国花园城。此外，维多利亚花园浸润西欧古典园林的威仪与尊荣，对称中轴广场、精美草坪和精美绝伦的园艺造型，洋溢英伦贵族生活气息。

Other Styles

其他风格

Quiet and Secluded
曲径通幽

Simple and Flexible
简练随性

Noble and Elegant
高贵典雅

KEYWORDS 关键词

Ecological and Natural
生态自然

Pleasant and Leisure
轻松怡人

Landscape Space
空间景观化

Other Styles
其他风格

Location: Kunming, Yunnan
Landscape Design: UC Landscape Architecture Design Group
Landscape Area: 85,540 m²
Green Coverage Ratio: 45%

项目地点：云南省昆明市
景观设计：深圳源创易景观设计有限公司
景观面积：85 540 m²
绿化率：45%

Sanctuary, AVIC Kunming
昆明中航 · 云玺大宅

FEATURES 项目亮点

Landscape design adopts New Asian Style and carefully designs every single landscape node in order to highlight the ecological and natural environmental design concept.

项目的景观设计采用现代“新亚洲风格”，对每个景观节点加以精心打造，突出生态、自然的环境设计理念。

Overview

Sanctuary is located outside of Second Ring Road, Guandu District, Kunming, and east bank of the Dian Lake, close to Wujiatang Wetland Park, which is the point of golden section of downtown and Chenggong New Town, while Guangfu Road is in the north, Huanghu East road in the south, and it is lying on Erji Road in the east and crossed by Changhong Road from north to south. It is approx. 10 km away from urban area, 6 km away from business circle of Century Town and New Asia.

The landscape axis of Dian Lake east bank is grand and magnificent, with clean and symmetrical landscape which has followed the features and atmosphere of New Asian Style. With this distinctive theme as background, plants in various colors working together with different pavements are featured as the distinctive items of the project to represent a landscape feast to residents.

项目概况

项目位于昆明市官渡区二环路外，滇池东岸，五甲塘湿地公园旁，属于昆明主城区及呈贡新城之间黄金分割点。北临广福路，南接环湖东路，西依五甲塘湿地公园，东靠珥季路，昌宏路南北贯穿，距离主城中心约 10 km，距世纪城—新亚洲商圈约 6 km。

滇池龙岸轴线景观尺度恢宏又隆重，轴线对称又明确，继承了新亚洲风格的气质和氛围。在这样主题鲜明的背景映衬下，各色植物、不同铺装形成了本项目独有的特色，尽情地向人们展示着一场景观盛宴。

SANCTUARY

Site Plan 总平面图

Courtyard Wall and Space Between Houses Section
院墙宅间断面图

Design Objective

Landscape design adopts New Asian Style and carefully designs every single landscape node in order to highlight the ecological and natural environmental design concept.

The purpose of landscape design is to not only create a residential harbor that is connecting closely with Kunming City, but also build an independent living environment and an innovative landmark to promote its brand value and the value in use furthest.

设计目标

项目的景观设计采用现代“新亚洲风格”，对每个景观节点加以精心打造，突出生态、自然的环境设计理念。

项目的景观设计宗旨是为昆明这个城市创造一个与之紧密相连的住宅港湾的同时，还要创造一个独立的居住空间环境和具有创新性的地标性场所，最大程度地提升其品牌价值及使用价值。

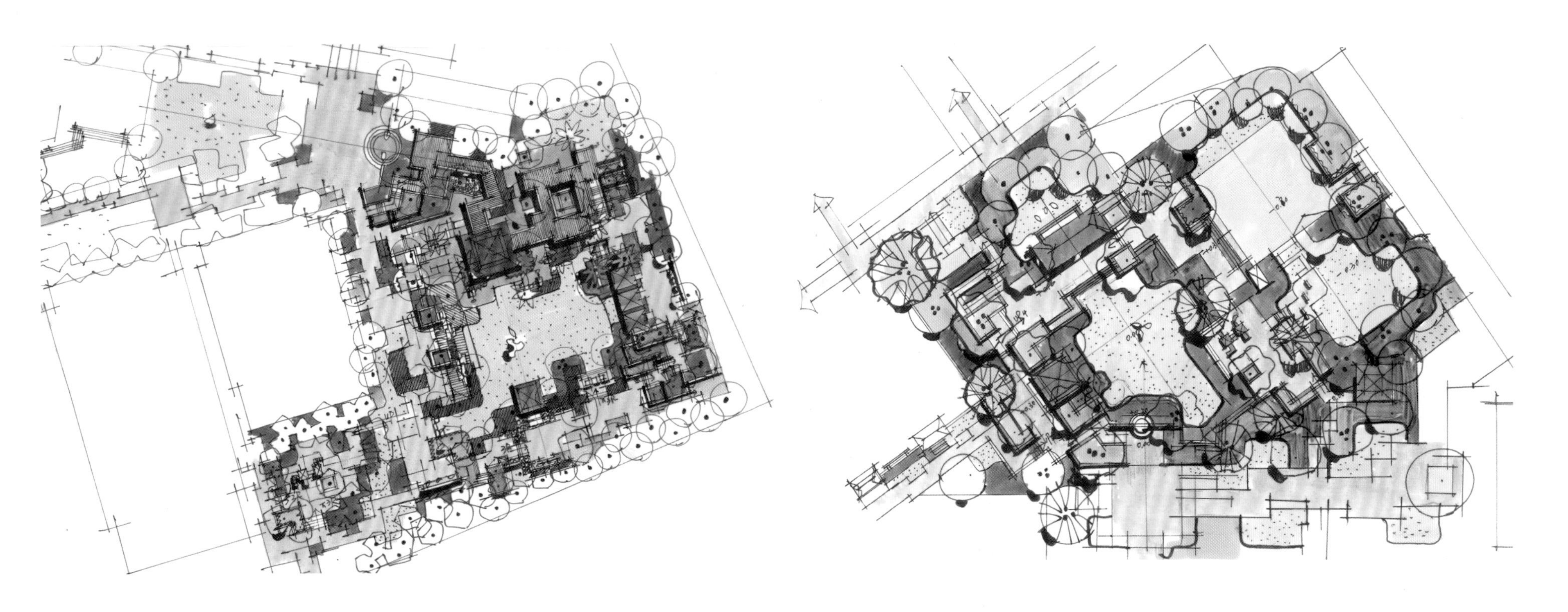

Landscape Planning

Landscape designers are eager to combine traditional Chinese residential ideas with modern life concepts together in this project. In this case, designers leave all the above ground area to residences and set underground garage for parking, while sunken courtyards are designed for landscaping underground space to lay the foundation for creating pleasant and leisure environment of the community.

Landscape designers have divided the walkways and roads in purpose to ensure the smooth of both pedestrian and vehicles. For the design strategy, designers have worked their best to reach the sustainability of living environment and to improve the living quality through effort on the ecological environment. Interpersonal interaction and open spaces have been the foundation of landscape overall planning.

景观规划

在这个项目里，设计师希望把中国的传统居住理想与现代的生活理念相结合。为此，设计师首先把地面还给居民人行使用，所有停车在地下室解决，而下沉庭院则把地下室空间景观化、地面化，从而为社区塑造轻松怡人的环境奠定了基础。

景观总体规划具有针对性的将公共交通与人行流线区别开来，以此保证人流和交通的顺畅。在设计策略中将居住环境的可持续发展开发到最大化，通过生态环境的过程提高居住质量。人际互动、开放空间构成了景观总体规划的根本。

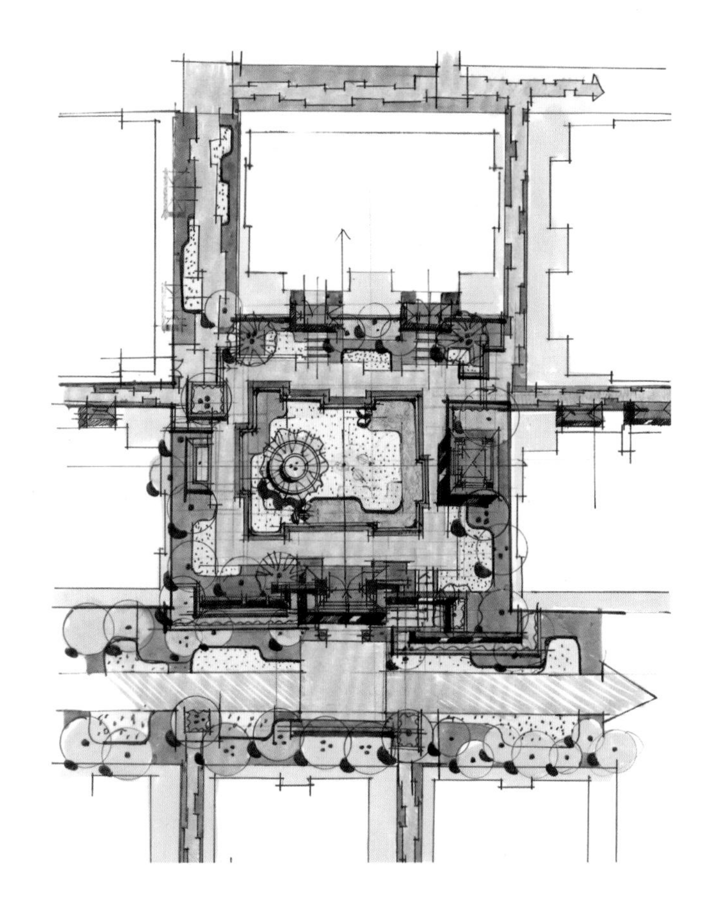

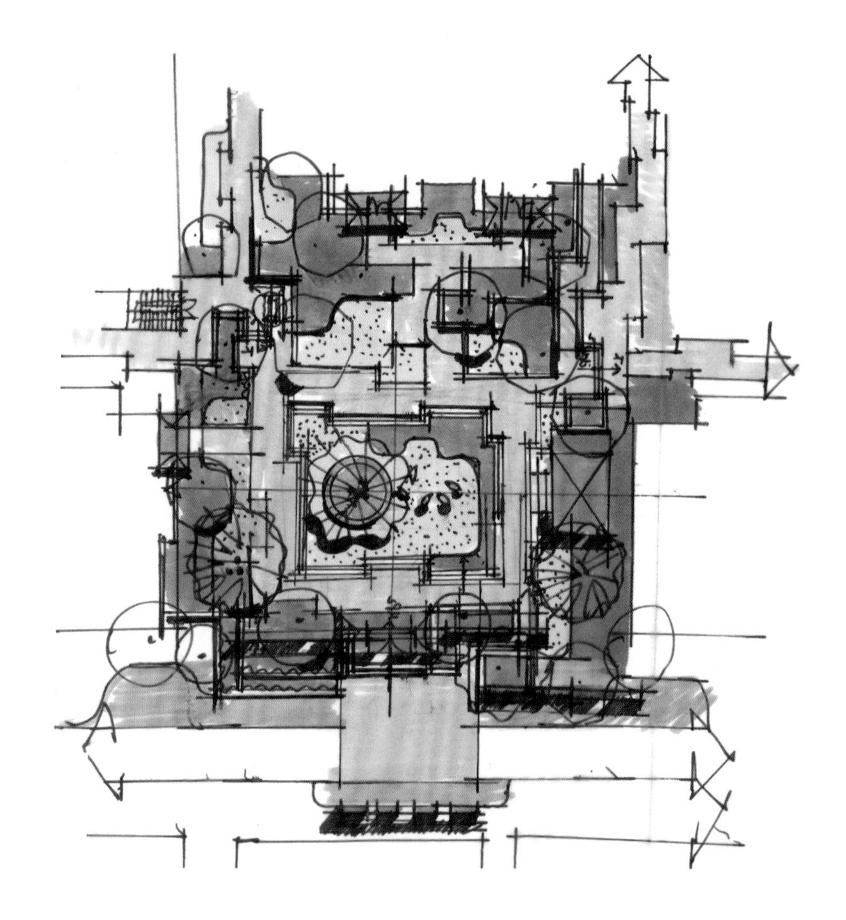

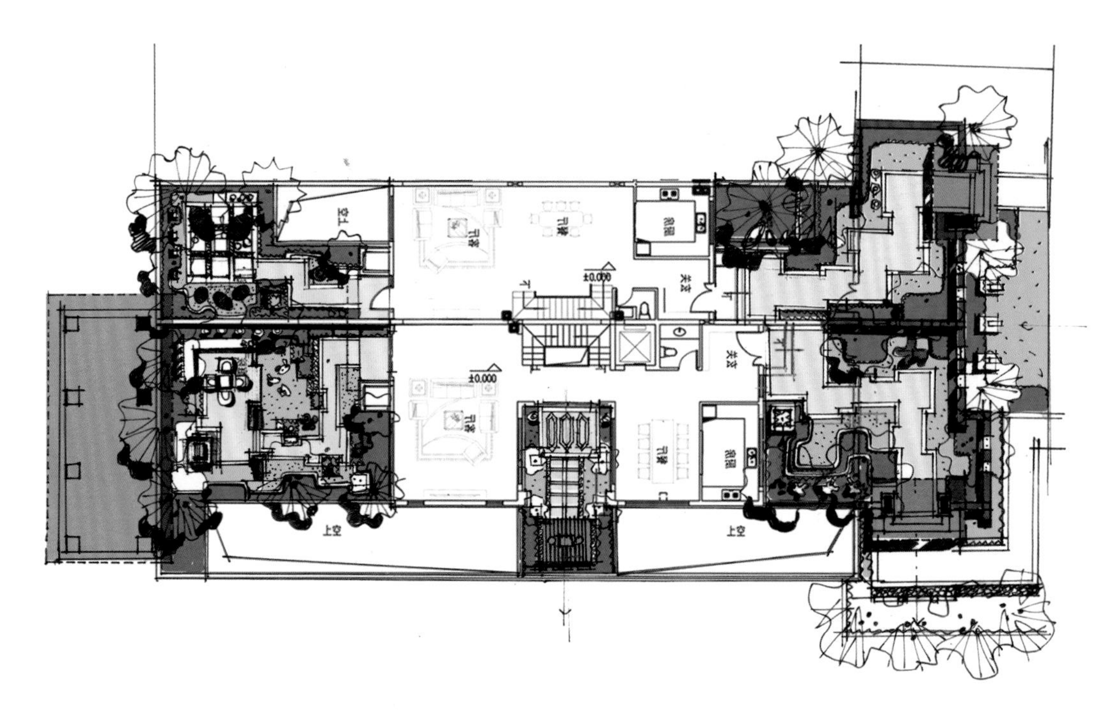
±0.000
±0.000
上空
上空

KEYWORDS 关键词

Borrowed Scenery
借景山水

Courtyard Atmosphere
庭院氛围

Adjusting to Local Conditions
因地制宜

Other Styles
其他风格

Location: Fuyang, Zhejiang
Developer: Zhongda (Fuyang) Real Estate Co., Ltd.
Architectural Planning: JWDA
Landscape Design: Palm Landscape Architecture Co., Ltd.
Gross Land Area: 421,805 m^2
Plot Ratio: 0.9

项目地点：浙江省富阳市
开 发 商：富阳中大房地产有限公司
规划设计：骏地建筑设计咨询有限公司
景观设计：棕榈设计有限公司
总占地面积：421 805 m^2
总容积率：0.9

Zhongda · Hangzhou West Peninsula
中大·杭州西郊半岛

FEATURES 项目亮点

The general layout of this project is full of mountains and water features, and the residential buildings that set into the hillside with irregular terrain are echoing with the overall environment to reach the harmony among mountain, river and human beings.

项目总体布局显山露水，群落依山而建，地势起伏不平，建筑感知环境，空间自然生长，达到“山、水、人”之间的和谐共存。

Overview

Developing as a diversified project collecting residence, resorts, transport, city revetment and public entertainment together, West Peninsula is acting as a part of city function and energy in the near future.

项目概况

杭州西郊半岛是集住宅、度假酒店、城市交通、城市驳岸、公共娱乐于一体的多元化项目，是城市未来发展中城市功能和活力的组成部分。

1 MAIN ENTRANCE WITH SPECIAL PAVING 主入口特色铺装
2 CLUB HOUSE MAIN ENTRANCE WITH SPECIAL PAVING 会所主入口特色铺
3 CLUB HOUSE TERRACE 会所活动平台
4 CLUB HOUSE SWIMMING POOL 会所游泳池
5 SPA SPA
6 WOOD DECK 木平台
7 PAVILLION 亭子
8 MAIN WATER FEATURE 主水景
9 POND WITH WATER FEATURE 景观池塘
10 OPEN LAWN AREA 草坪活动空间
11 SCREENING EVERGREEN ON MOUND 常绿植物
12 OVERLOOK 眺望台
13 TOWER ENTRANCE WITH SPECIAL PAVING 塔楼入口特色铺装
14 CHILDREN'S PLAYGROUND 儿童游乐场
15 EMERGENCY VEHICLE ENTRANCE 紧急车辆出入口
16 SURFACE PARKING 路面停车

Site Plan 总平面图

Design Objective

Fuyang is known throughout the world by Huang Gongwang's painting *Dwelling in the Fuchun Mountains*, the design concept of the development is inspired by the study and analysis on the local environment and culture to adjust each plot cooperating with its own conditions and also combining with several architectural style, creating the landscape layer by layer and borrowing the original landscape of mountain and river of Fuyang to make the landscape integrate with the whole nature.

设计目标

富阳以黄公望的《富春山居图》而闻名于世，通过对区域环境及地域文化的深入分析和提炼，设计理念得以启发，每个地块因地制宜，融合不同的建筑风格，阶梯化设计景观，充分借景富春山水，力求做到景观和整个大自然融为一体。

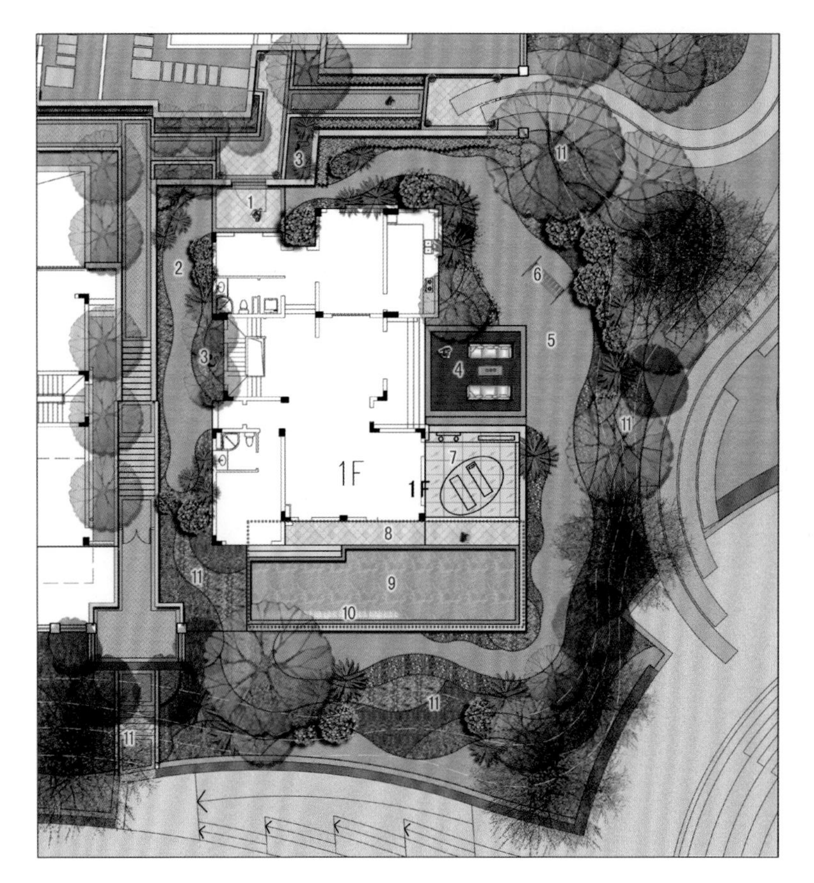

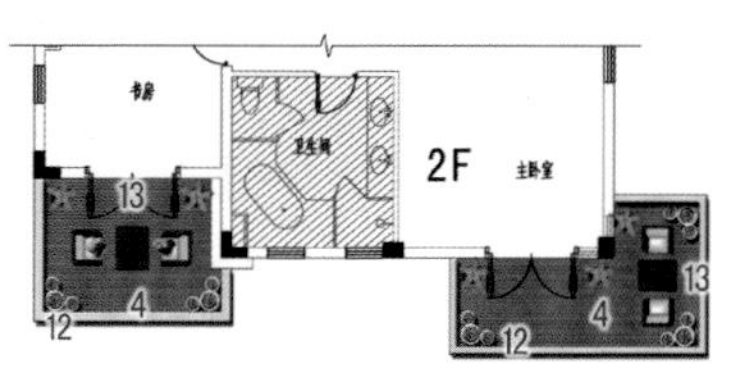

二层露台景观平面图

图例:

1 入户平台
2 庭院绿化
3 景观小雕塑
4 休息木平台
5 阳光草坪
6 庭院秋千
7 玻璃阳光房
8 下沉庭院平台
9 下沉庭院泳池
10 下沉庭院水景
11 自然种植区
12 陶罐花钵
13 休憩沙发
14 下沉庭院景墙

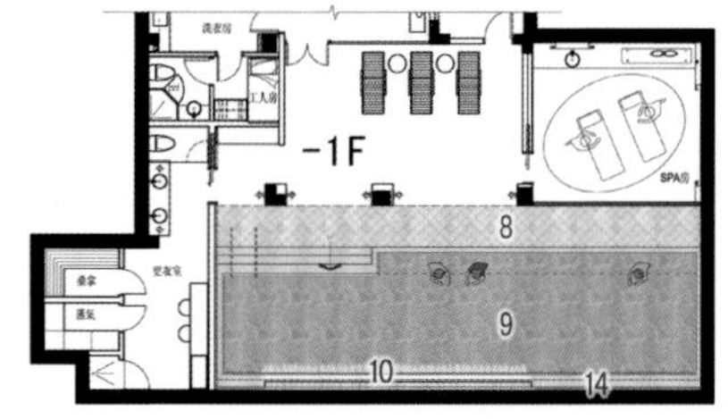

负一层景观平面图

Landscape Planning

Following on the characteristics of original environment and meeting the requirements of overall planning, architectural features and project orientation, designers make the axis and landscape nodes echoing and connecting with each other to extend the view, also they are people oriented, and trying to improve the user experience on every single corner. A great amount of landscaping methods, such as obstructive and opposite scenery, borrowed scenery as well are used for this project to make each landscape nodes cooperate with the others and to create multi-layered courtyard atmosphere.

景观规划

设计师根据实际的环境特点，结合整体规划、建筑特征、项目定位等要求，轴线与景观点互相呼应、关联，寻找视线的延伸线，以人为本，提升每一个空间的用户体验。设计大量运用障景、对景、借景等造景手法，让项目的各个节点互相依托，多层次地营造出居住区的庭院氛围。

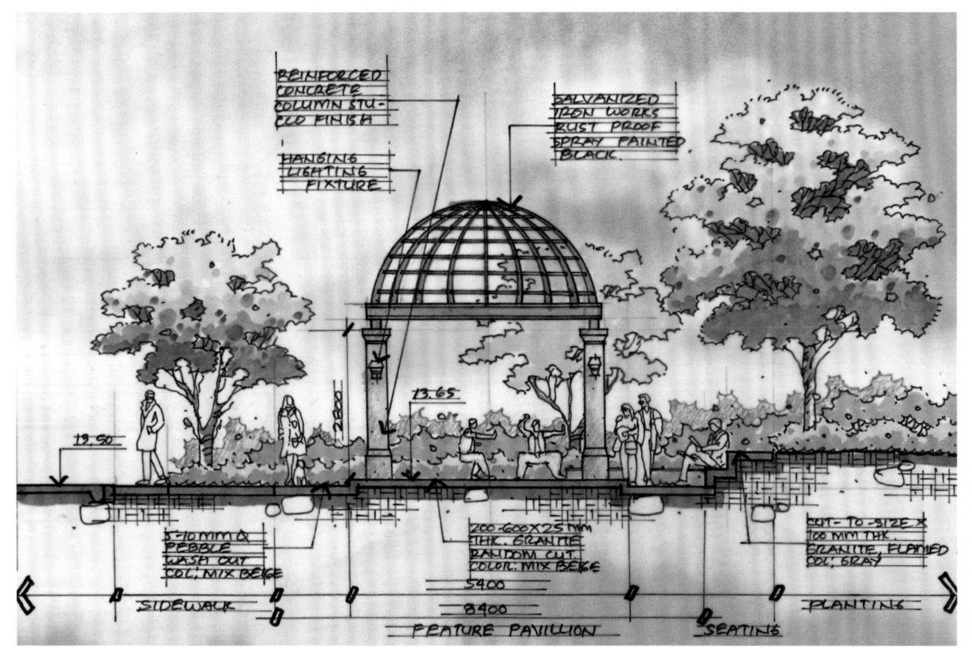

Clubhouse Area Swimming Pool Section
会所区泳池剖面图

Concentrated Green Area Section
集中绿地区剖面图

Clubhouse Area Swimming Pool Section
会所区泳池剖面图

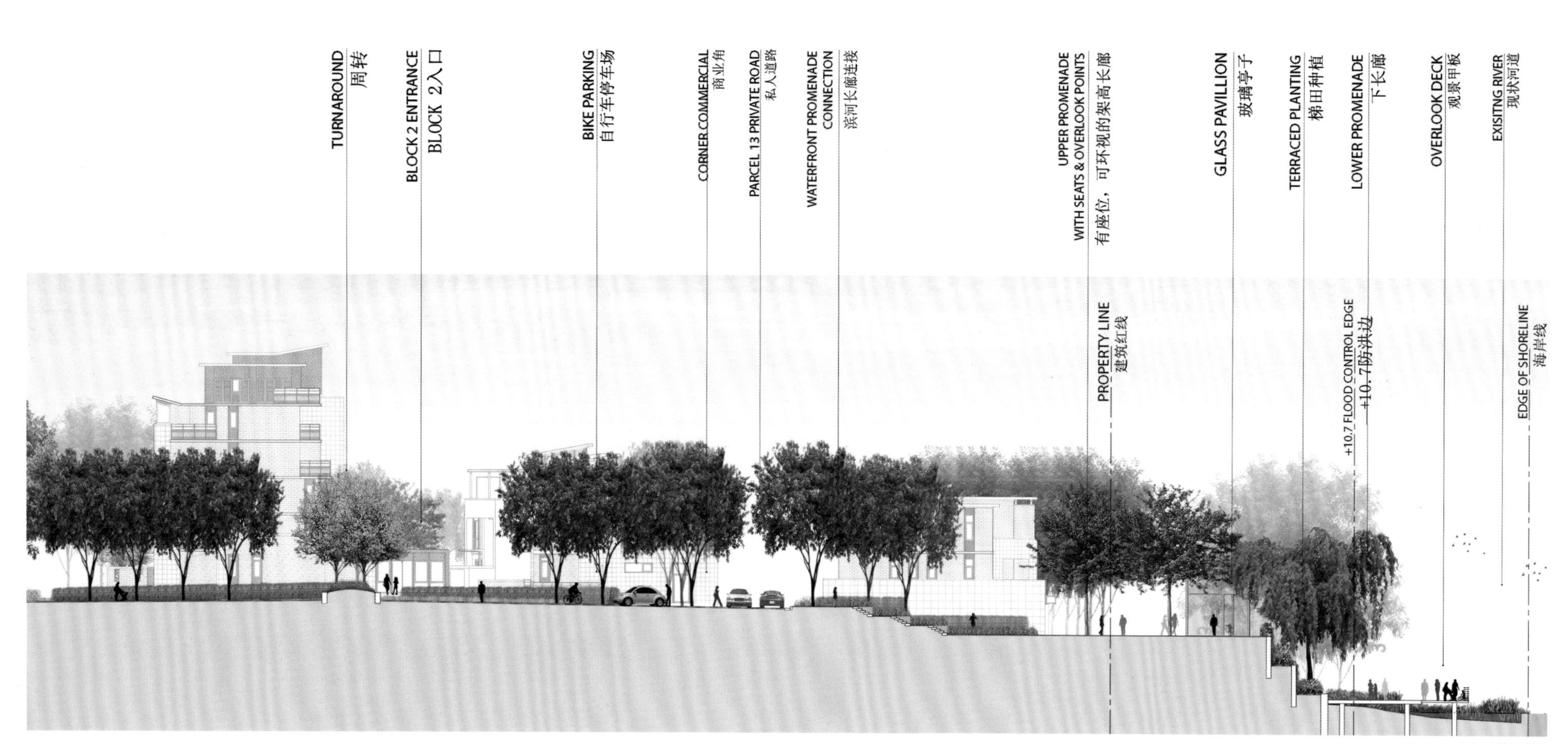

Section 1 剖面图 1

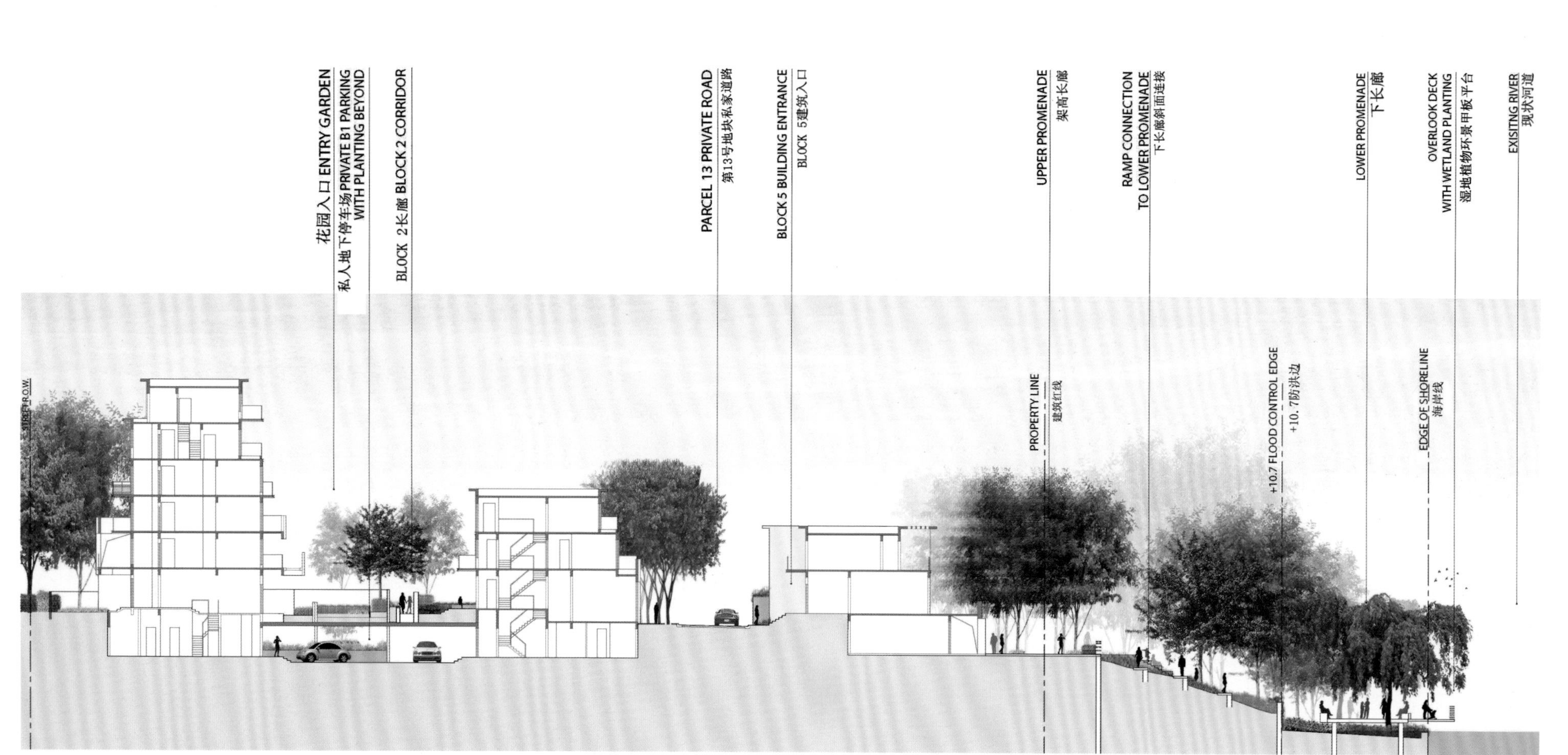

Section 2 剖面图 2

KEYWORDS 关键词

Gingko Avenue
银杏大道

Five-layer Landscape
五重绿化

Grand Atmosphere
气势宏大

Other Styles
其他风格

Location: Jialing District, Nanchong, Sichuan
Landscape Design: Hong Kong Merily International (Chengdu) Co., Ltd.
Land Area: 222,111 m²

项目地点：四川省南充市嘉陵区
景观设计：香港美林国际（成都）景观设计有限公司
占地面积：222 111 m²

Tianlu Villa, Nanchong
南充天庐别墅

FEATURES 项目亮点

With the highlight of unique and closeness of landscape, it includes exclusive five-layer landscape, enclosed community with 2.5 km gingko landscape avenue, Southern California flavor gardens and well-placed flowing water park.

项目在景观设计上强调独创和亲和。项目拥有独有的五重景观绿化、2.5 km的银杏景观大道围合社区、南加州风情的园林以及鳞次栉比的活水公园。

Overview

Located at south end of Coastal Avenue, Jialing District, Nanchong City, Tianlu Villa is lying on the flourishing Fengya Mountain and facing to green Jialing River, connecting various plots together and occupying a landmark residence area of Nanchong City. It adopts North American country style and will be the largest high-end community of Nanchong City. The plot ratio of villas is 0.4 and that of multi-storey residences is 1.0, while independent villas close to the river are the most important selling points of the project.

项目概况

天庐别墅位于南充市嘉陵区滨江大道南端，背靠青山如黛的凤垭山，前邻碧波荡漾的嘉陵江，连接千年绸都万亩桑莘，占据了"成渝第三城"居住里程碑的人文地脉。项目采用北美乡村风格，建成后将成为南充市最大的高端生活居住区。其中，天庐别墅区容积率 0.4，洋房区容积率 1.0，亲水独栋别墅的更是项目的主要卖点。

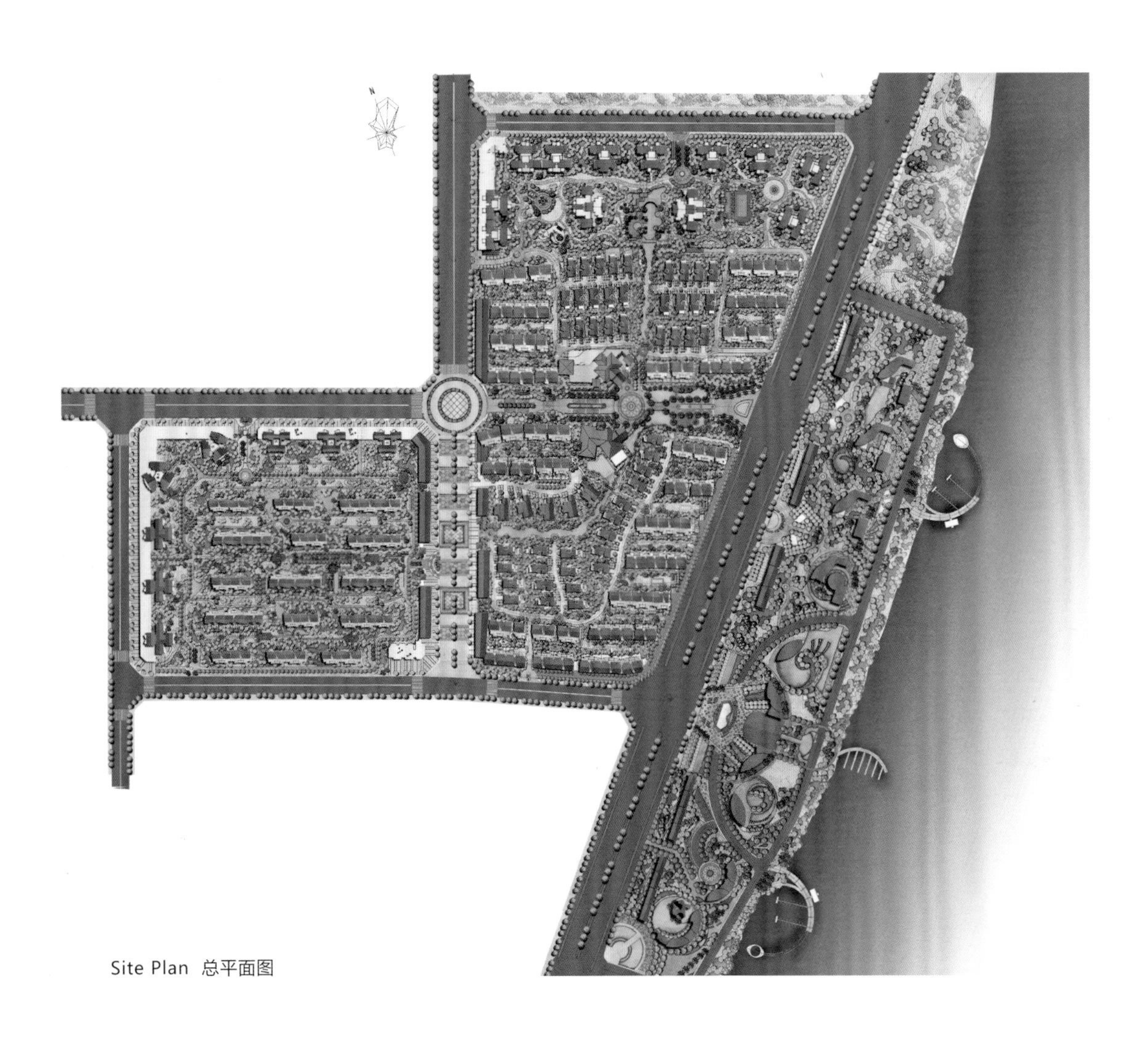

Site Plan 总平面图

Design Objective

Tianlu is planning to build a five-star hotel and a thematic water park as VIP facilities and regional function facilities. By adopting the constructing methods, crafts and materials of villas to the whole project even every small corner, designers are eager to create distinguished, beautiful and comfortable living environment.

设计目标

"天庐"立意高远，领袖南充、比肩成渝，项目规划了一所五星级酒店和一个情景水上游乐园作为项目VIP配套及城市区域功能配套。同时坚持以别墅的营造手法、工艺和材质贯穿项目开发的大小角落和始终，铸造尊贵、唯美、舒适的居住华章。

Landscape Planning

With the highlight of unique and closeness of landscape, it includes exclusive five-layer landscape, enclosed community with 2.5 km gingko landscape avenue, Southern California flavor gardens and well-placed flowing water park. Moreover, the square is large and grand. A large sculpture is standing in the front pool with fountains on the two sides, and a lot of tropical plants are also planted around.

景观规划

项目在景观设计上强调独创和亲和。项目拥有独有的五重景观绿化、2.5 km 的银杏景观大道围合社区、南加州风情的园林以及鳞次栉比的活水公园。广场以宏大的气派示人，其正前方矗立有巨大的艺术雕塑于水池中，两边都设有喷泉，四周栽种有诸多的热带植物。

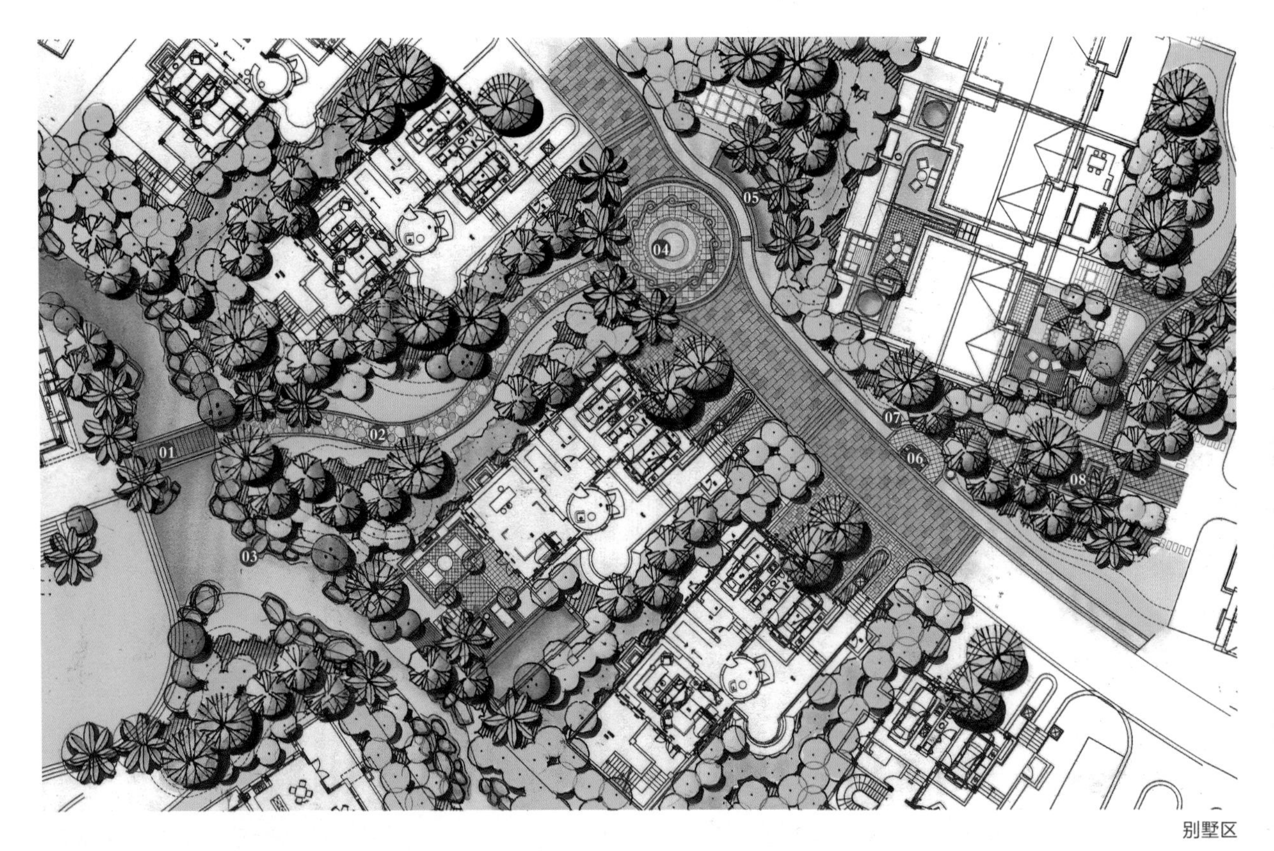

别墅区

GRAND CANYON
NATIONAL PARK

KEYWORDS 关键词

Winding & Quiet
曲折宁静

Waterfront Valley
水边坡谷

Group Division
组团分割

Other Styles
其他风格

Location: Shenzhen, Guangdong
Landscape Design: Baroque Design Group Co., Ltd
Land Area: 200,322 m^2

项目地点：广东省深圳市
景观设计：香港博唯规划建筑与环境景观设计有限公司
占地面积：200 322 m^2

Galaxy Dante
星河丹堤

FEATURES 项目亮点

The spacious lake makes people lively and cheerful, and the babbling brook offers people winding and quiet sense.

广阔的湖面使人明快开朗，潺潺的流水，更让人感到曲折宁静。

Overview

The Galaxy Dante, located at the north of Caitian Road, west of Yinhu Mountain, the junction of Meiguan Highway and Nanping Express, is near the urban downtown. It's surrounded by mountains and lakes with fantastic environment and natural landscape. It's with a gross land area of 200,000 m^2, gross floor area of 360,000 m^2 and plot ratio of 1.8.

项目概况

星河丹堤位于彩田路北、银湖山西，地处梅关高速与南坪快速交界处，临近城市中心。山湖环绕，风水绝佳，美景天成。项目总用地面积为 20 万 m^2，总建筑面积为 36 万 m^2，容积率达 1.8。

星河丹堤

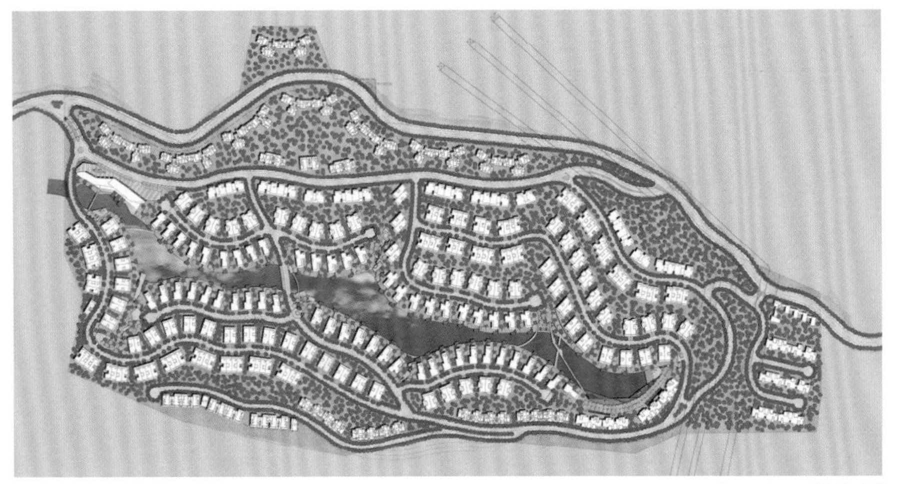

Site Plan 总平面图

Design Objective

The Galaxy Dante strives to provide the residents with an ecological, organic and environmental living environment under the strict planning structure and full integration of natural landscape. It creates a people-oriented and nature-integrated living space. It uses modern design and embellishment to create a rich and pleasing landscape. It uses modern design to embellish and create a rich and pleasant landscape, comprehensively analyzes the landscape spatial form and strives to penetrate the layout and geographic environment for harmonious coexistence.

设计目标

星河丹堤致力于在严谨的规划结构下，与自然景观充分融合，为住户提供一个生态、有机和环保的生活环境，营造以人为本、天人合一的生活场域。现代设计的修饰打造丰富的怡人景观，全面分析景观空间形式、布局与地理环境，力求相互渗透，协调共生。

Landscape Planning

The planning is from west to east and all is surrounded by lakes and mountains. The spacious lake makes people lively and cheerful. The babbling brook offers people a winding and quiet sense. The designer imitates the "end of the road" design of the American suburban residence, which can ensure the harmonious neighborhood relationships and the privacy of each unit, as well as show the harmony of landscape and residence. The innovation and application of new pacific style are full of modern sense and remain the beauty of nature.

The designer considers the use and creation of slopy terrain, and plans drainage at lowland, which has the function of water draining and makes the waterfront valley become the natural background of this project. The natural landscapes such as Eco green valley, mountain and bay mouth divide the community into several groups. The small high-rise is divided into 2 groups to protect the mountain landscape, visual corridor and the mountain.

景观规划

规划自西向东，环湖傍山。广阔的湖面使人明快开朗，潺潺的流水，更让人感到曲折宁静。设计师仿效美国郊区住宅的“尽端路”设计，既要保证邻里关系的共融及各单元的私密性，又能体现景观与住宅的共融。而新太平洋风格的革新应用，富含现代感之余又保留着自然的美感。

设计师同时考虑到坡地形态的利用营造，并在低地规划水系，既有地面排水功效，同时水边坡谷又成为该项目的自然背景。生态绿谷、山体、湾口等自然景观将社区分割成若干组团村落。尤其是小高层分成 2 个组团，有利于保护山景、视觉通廊及山体。